Quality of Service in Optical Burst Switched Networks

OPTICAL NETWORKS SERIES

Series Editor
Biswanath Mukherjee, *University of California, Davis*

Other books in the series:

OPTICAL WDM NETWORKS
Biswanath Mukherjee, ISBN 0-387-29055-9
TRAFFIC GROOMING IN OPTICAL WDM MESH NETWORKS
Keyao Zhu, Hongyue Zhu, Biswanath Mukherjee, ISBN 0-387-25432-3
SURVIVABLE OPTICAL WDM NETWORKS
Canhui (Sam) Ou and Biswanath Mukherjee, ISBN 0-387-24498-0
OPTICAL BURST SWITCHED NETWORKS
Jason P. Jue and Vinod M. Vokkarane, ISBN 0-387-23756-9

QUALITY OF SERVICE IN OPTICAL BURST SWITCHED NETWORKS

KEE CHIANG CHUA
MOHAN GURUSAMY
YONG LIU
MINH HOANG PHUNG
National University of Singapore

Kee Chaing Chua
Mohan Gurusamy
Yong Liu
Minh Hoang Phung
Department of Electrical and Computer Engineering
National University of Singapore
Singapore

Quality of Service in Optical Burst Switched Networks

e-ISBN 0-387-47647-6

ISBN 978-1-4419-4164-0 e-ISBN 978-0-387-47647-6

Printed on acid-free paper.

Printed in the United States of America.

9 8 7 6 5 4 3 2 1

springer.com

Specially dedicated to:
Nancy, Daryl and Kevin Chua

-Kee Chaing Chua

My parents and wife

-Mohan Gurusamy

My parents and wife

-Yong Liu

My family

-Minh Hoang Phung

Contents

Preface

Optical Burst Switching (OBS) is a promising switching architecture to support huge bandwidth demand in optical backbone networks that use Wavelength Division Multiplexing (WDM) technology. Due to its special features which combine the merits of optical circuit switching and packet switching, it can support high-speed transmission with fine bandwidth granularity using off-the-shelf technologies. OBS has attracted a lot of attention from researchers in the optical networking community. This book is devoted to a comprehensive discussion of the issues related to supporting quality of service (QoS) in OBS networks. Some of these issues include various mechanisms for providing QoS support to multiple traffic classes including absolute as well as relative differentiation frameworks, edge-to-edge QoS provisioning and other non-mainstream variations of mechanisms that have been reported in recent literature. It is hoped that this work will provide individuals interested in QoS provisioning in OBS networks with a comprehensive overview of current research and a view of possible directions for future research.

Singapore,
June 2006

Kee Chaing Chua
Mohan Gurusamy
Yong Liu
Minh Hoang Phung

1

INTRODUCTION

1.1 Evolution of Optical Networks

Since the advent of the World Wide Web, the Internet has experienced tremendous growth. Everyday, more and more people turn to the Internet for their information, communication and entertainment needs. New types of applications and services such as web browsing, video conferencing, interactive online gaming, and peer-to-peer file sharing continue to be created to satisfy these needs. They demand increasingly higher transmission capacity from the networks. This rapid expansion of the Internet will seriously test the limits of current computer and telecommunication networks. There is an immediate need for new high-capacity networks that are capable of supporting these growing bandwidth requirements.

Wavelength Division Multiplexing (WDM) [1, 2] has emerged as a core transmission technology for next-generation Internet backbone networks. It provides enormous bandwidth at the physical layer with its ability to support hundreds of wavelength channels in a single fibre. Systems with transmission capacities of several Terabits per second have been reported [3]. In order to make efficient use of this raw bandwidth, efficient higher layer transport architectures and protocols are needed.

First-generation WDM systems, which are deployed in current backbone networks, comprise WDM point-to-point links. In these networks, routers are connected by high-bandwidth WDM links. At each router, all incoming Internet Protocol (IP) packets are converted from the optical domain to the electronic domain for processing. At the output links, all outgoing packets are converted back from the electronic domain to the optical domain before being transmitted on outgoing fibres. Since the electronic processing speed is much lower than the optical transmission rate, opto-electronic-opto (O-E-O) conversion of the entire traffic stream at every router creates significant overheads for the system, especially when most of the traffic is by-pass traffic.

Optical networking has become possible with the development of three key optical network elements: Optical Line Terminator (OLT), Optical Add/Drop Multiplexer (OADM) and Optical Cross Connector (OXC) [4]. An OLT multiplexes multiple wavelengths into a single fibre and demultiplexes a composite optical signal that consists of multiple wavelengths from a single fibre into separate fibres. An OADM is a device that takes in a composite optical signal that consists of multiple wavelengths and selectively drops (and subsequently adds) some of the wavelengths before letting the composite signal out of the output port. An OXC has multiple input and output ports. In addition to add/drop capability, it can also switch a wavelength from any input port to any output port. Both OADM and OXC may have wavelength conversion capability. These devices make it possible to switch data entirely in the optical domain between a pair of source and destination nodes.

The next-generation optical Internet architecture is envisioned to have two main functional parts: an inner core network and multiple access networks [5]. The access networks are compatible with today's Internet transport architecture and are responsible for collecting IP traffic from end-users. They are built from electronic or lower-speed optical transport technologies such as Gigabit Ethernet or optical rings or passive optical networks (PONs). The access networks are connected together by the inner core network through high-speed edge nodes. An ingress node aggregates traffic destined to the same egress node and forwards it through the core network. The core network consists of a mesh of reconfigurable optical switching network elements (e.g., OXC and OADM) interconnected by very high capacity long-haul optical links. To date, there are primarily three all-optical transport technologies proposed for the optical core network, namely wavelength routing [6] or Optical Circuit Switching (OCS), Optical Packet Switching (OPS) and Optical Burst Switching (OBS). They are described below.

In OCS networks, dedicated WDM channels, or lightpaths, are established between a source and destination pair. The lightpath establishment may be static or dynamic. A lightpath is carried over a wavelength on each intermediate link along a physical route and switched from one link to another at each intermediate node. If wavelength converters are present in the network, a lightpath may be converted from one wavelength to another wavelength along the route. Otherwise, it must use the same wavelength on all the links along the route. This property is known as the wavelength continuity constraint. A wavelength may be used by different lightpaths as long as they do not share any common link. This allows a wavelength to be reused spatially at different parts of the network.

Although the wavelength routing approach is a significant improvement over the first generation point-to-point architectures, it has some limitations. Firstly, lightpaths are fairly static and fixed-bandwidth connections that may not be able to efficiently accommodate the highly variable and bursty Internet traffic. In addition, the number of connections in a network is usually much greater than the number of wavelengths and the transmission rate of a connection is much smaller than the capacity of a wavelength. Therefore, despite spatial reuse of wavelengths, it is neither possible nor efficient to allocate one wavelength to every connection. This problem can be alleviated by traffic grooming [7, 8], which aggregates several connections into a lightpath. However, some connections must still take multiple lightpaths when there is no single lightpath between a pair of source and destination nodes. Such connections will have to undergo multiple O-E-O conversions and multiple crossings through the network, increasing network resource consumption and core network edge-to-edge delay.

OPS [9, 10, 11, 12, 13] is an optical networking paradigm that performs packet switching in the optical domain. In this approach, optical packets are sent along with their headers into the network without any prior reservation or setup. Upon reaching a core node, a packet will be optically buffered while its header is extracted and processed electronically. A connection between the input port and the output port is then set up for transmission of that optical packet and the connection is released immediately afterwards. As such, a link can be statistically shared among many connections at subwavelength level. OPS may have slotted/unslotted and synchronous/asynchronous variants.

The objective of OPS is to enable packet switching capabilities at rates comparable with those of optical links and thereby replacing wavelength routing in next-generation optical networks. However, it faces several challenges involving optical technologies that are still immature and expensive. One such challenge is the lack of optical random access memory for buffering. Current optical buffers are realized by simple Fiber Delay Lines (FDLs), not fully functional memories. Other required technologies that are still at a primitive stage of development include fast optical switching, optical synchronization and the extraction of headers from optical packets.

OBS [14, 15, 16, 18, 19] is a more recently proposed alternative to OPS. In OBS, the basic transport unit is a *burst*, which is assembled from several IP packets at an ingress node. OBS also employs a one-pass reservation mechanism, whereby a header packet is sent first to reserve wavelengths and configure the switches along a path. The corresponding burst follows without waiting for an acknowledgement for connection establishment. If a switch along the path cannot forward the burst due to contention, the burst is simply dropped. This mechanism has its origin in an ITU-T standard for Asynchronous Transfer Mode (ATM) networks known as ATM Block Transfer with Immediate Transmission (ABT-IT) [20]. Other variants of ABT-IT include Ready-to-Go Virtual Circuit Protocol (RGVC) [21] and Tell-and-Go (TAG) [22]. The use of large bursts as the basic transport unit leads to lower switching frequency and overhead. Therefore, OBS nodes can use slower switching fabrics and processing electronics compared to OPS. The overhead reduction occurs in two places. Firstly, the header/payload ratio is reduced, leading to lower signaling overhead. Secondly, the

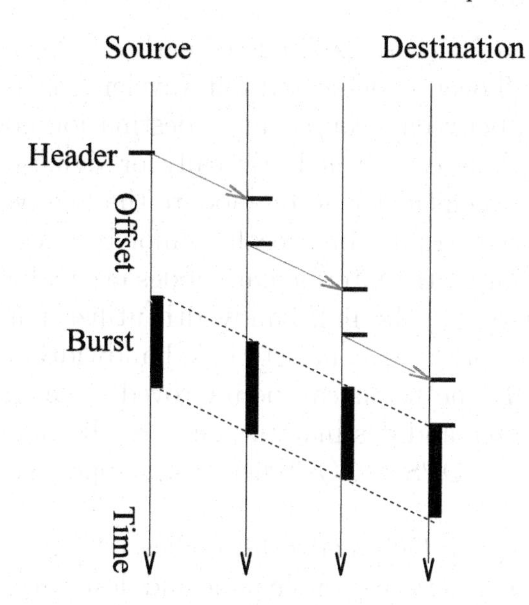

Fig. 1.1. The use of offset time in OBS

ratio between guard intervals between bursts when a link is idle and the time it is transmitting is also reduced.

A distinguishing feature of OBS is the separation between a header packet and its data burst in both time and space. A burst is not sent immediately after the header packet, but delayed by a predetermined *offset* time. The offset time is chosen to be at least equal to the sum of the header processing delays (δ) at all intermediate nodes. This is to ensure that there is enough time for each node to complete the processing of the header before the burst arrives. The use of the offset time is illustrated in Figure 1.1. Header packets are transmitted in dedicated control channels, which are separate from data channels. This separation permits electronic implementation of the signaling control path while maintaining a completely transparent optical data path for high speed data transmission. It also removes the need for optical buffering, optical synchronization and optical header extraction techniques.

Table 1.1 summarises the three all-optical transport paradigms. From the table, one can observe that OBS has the advantages of both OCS and OPS while overcoming their shortcomings.

- *Bandwidth utilization:* As discussed earlier, a lightpath in OCS networks will occupy dedicated full wavelengths along the end-to-end path between a source and a destination node. Whether the lightpath is established statically or dynamically, it will not be able to efficiently accommodate the highly variable and bursty Internet traffic. Lightpath cannot be used by the cross traffic starting at the intermediate nodes even when it is lightly loaded. Therefore, the link bandwidth utilization is quite low in OCS networks. In contrast, link bandwidth utilization in OPS and OBS networks can be improved since traffic between different source and destination pairs are allowed to share link bandwidth, i.e. OPS and OBS networks support statistical multiplexing.
- *Setup latency:* In OCS networks, dedicated signaling messages need to be sent between source node and destination node to set up and tear down a lightpath. Therefore, the setup latency is considered to be high as compared to OPS and OBS networks, which require only one-way signaling prior to data transfer.
- *Switching speed:* In OCS networks, a switching speed required is slow since the switching entity is lightpath which has relatively longer duration. The switches in OCS networks thus have enough time for dynamic configuration. However, in OPS networks, switches need to switch incoming optical packets quickly to different ports upon their arrival. Therefore, fast switching capability and reservation are required for switches in OPS networks. However, because of large granularity of bursts and offset time, the switching configuration time in OBS networks need not be as long as OPS switches. But, as compared to OCS, OBS needs fast switching fabric.
- *Processing complexity:* Since in OCS networks a lightpath has relatively longer duration, the complexity of OCS is relatively low when compared with OPS and OBS networks. However, since in OPS networks, the switching entity is individual optical packet, the complexity of OPS will be quite high. In OBS networks, the switching entity is individual data burst which is assembled by multiple individual packets. Therefore, the complexity of OBS is between that of OCS and OPS.

Table 1.1. Comparison of the different optical switching paradigms.

Optical Switching Paradigm	Bandwidth Utilization	Setup Latency	Switching Speed	Processing Complexity	Traffic Adaptivity
OCS	Low	High	Slow	Low	Low
OPS	High	NA	Fast	High	High
OBS	High	NA	Medium	Medium	High

- *Traffic adaptivity:* OCS networks cannot adapt well to support bursty traffic since the setup latency for a lightpath is quite high. However, OPS and OBS networks can adapt well to support bursty traffic due to traffic multiplexing.

1.2 Overview of OBS Architecture

1.2.1 System architecture

Figure 1.2 shows an OBS network. It comprises a meshed network of core nodes linked by WDM links. In the present literature, the OBS core nodes are usually assumed to have full wavelength conversion capability [15, 19]. In addition, depending on the switch architecture and design choice, the core nodes may or may not be equipped with optical buffering, which is in the form of FDL. However, FDL only offers short delay and cannot be considered as fully functional memory. Some core nodes also act as edge nodes, which means that they are connected to some access networks and accept IP input traffic as well as all-optical transit traffic. Depending on whether an edge node acts as a source or a destination, it can be called an ingress node or egress node, respectively.

Data bursts are assembled from input traffic by ingress nodes before being sent over the OBS core network. The ingress node architecture is shown in Figure 1.3. Data bursts are put into different queues to support differentiated QoS. A burst scheduling unit selects the next burst for transmission according to a burst scheduling algorithm. An offset time setting unit sets the offset time for each outgoing burst. When a burst is ready for transmission, the ingress node sends a header packet towards the egress

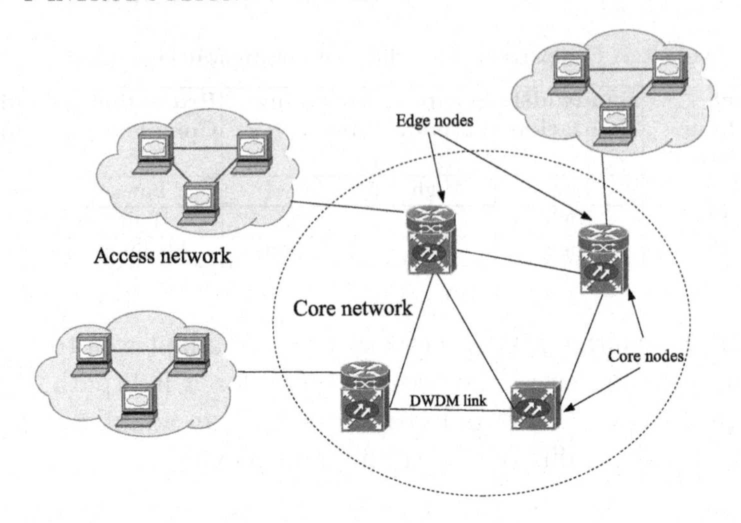

Fig. 1.2. OBS network architecture

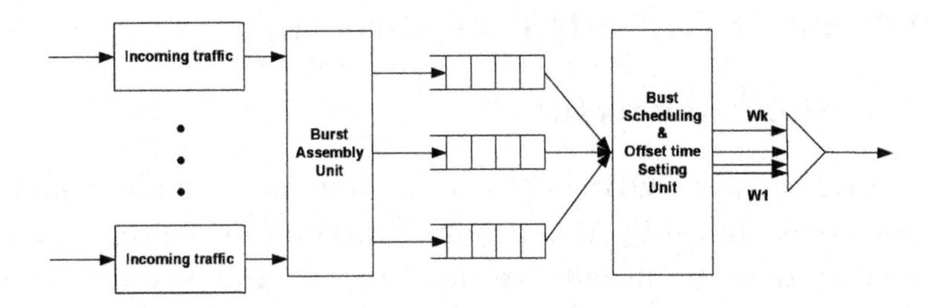

Fig. 1.3. Ingress node architecture in OBS networks

node on a dedicated control channel. The header packet carries information about the arrival time and size of the data burst which will be used to reserve wavelengths and configure switches at the ingress node and core nodes along the path. Then, the data burst is transmitted all-optically after its offset time without waiting for a connection setup acknowledgment. The core node architecture is shown in Figure 1.4. Each core node uses a burst scheduling algorithm to reserve wavelengths for data bursts. FDLs are used to hold data bursts to resolve wavelength contentions. Bursts are disassembled back into IP packets at egress nodes and forwarded onto adjacent access networks.

In the node architecture shown in Figure 1.4, a large number of tunable wavelength converters (TWCs) are required to achieve a desirable burst loss rate. To reduce the number of TWCs, an

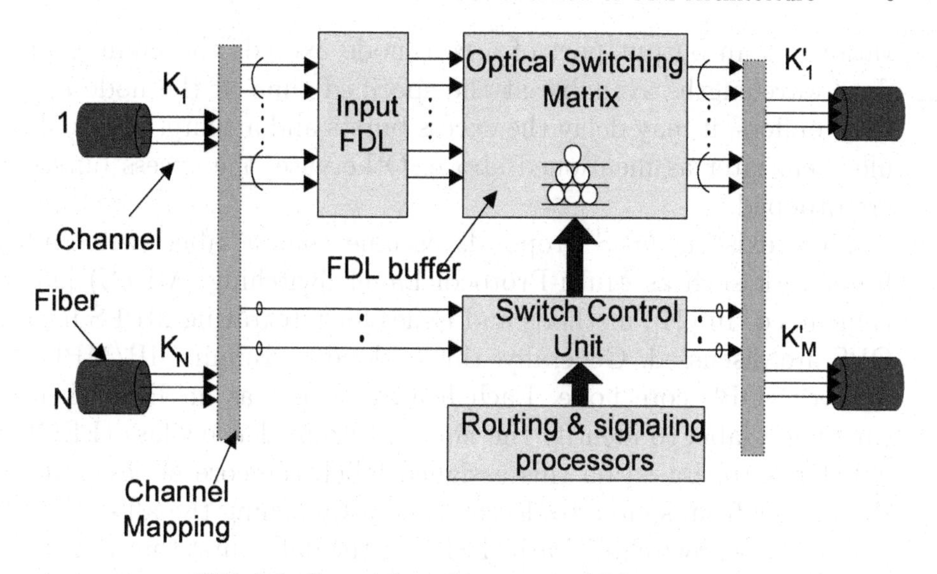

Fig. 1.4. Core node architecture in OBS networks

alternative node architecture which is called waveband-selective switching is proposed in [17] for OBS networks. In this switching architecture, W wavelengths are divided into K wavelength groups, which are referred as wavebands, each having the same number of continuous wavelengths (W/K). In edge nodes, each burst is also divided into W/K segments and is transmitted simultaneously on W/K wavelengths on any of the K wavebands. In core nodes, a burst will be switched to an available waveband using a tunable waveband converter (TWBC) instead of TWC. The benefit is that the number of TWBCs required for each input fiber port in this architecture is K which is much less than the number of TWCs required in Figure 1.4 which is W.

The details of OBS network architecture can be found in a related book [18], which is on the general architecture of Optical Burst Switched (OBS) networks while our book focuses specifically on Quality of Service (QoS) mechanisms.

Due to the limited buffer or bufferless nature of OBS networks, burst loss is the main performance metric of interest. Any burst queueing and assembly delay is confined to edge nodes, making it easy to manage. The primary cause of burst loss is burst contention. This happens when the number of overlapping burst reservations at an output port of a core node exceeds the number of data wavelengths available at the specified time. If the node has FDL buffers, it may delay the excess bursts and attempt to schedule them later as mentioned above. Otherwise, the excess bursts are dropped.

Most existing OBS proposals assume that a label switching framework such as Multi-Protocol Label Switching (MPLS) [23] is included. In [24], methods and issues for integrating MPLS into OBS are discussed. Generally, this is done by running IP/MPLS on every OBS core node. Each header is sent as an IP packet, carrying a label to identify the Forward Equivalence Class (FEC) it belongs to. Based on this assigned label, the core nodes route the header from source to destination, establishing the all-optical path or Label Switching Path (LSP) for the data burst that follows later. Label switching is the preferred routing method in OBS instead of hop-by-hop routing since its short label processing time per hop is particularly suitable for the high burst rate in OBS networks. Besides, label switching offers the possibility of explicit path selection, which enables traffic engineering.

1.2.2 Burst assembly mechanisms

Ingress edge nodes in OBS networks collect packets and assemble bursts with the same destination egress edge nodes. The burst is the basic transmission and switching unit inside OBS networks. Generally, assembly methods can be classified as timer-based and threshold-based. In a timer-based scheme, a timer is associated with the assembly process for each burst. When the timer expires, the edge node will stop the collection of packets and the burst is ready for transmission. In a threshold-based scheme, there will be an upper limit/threshold on the burst size for each burst. The edge node will stop the assembly process when the upper threshold is reached. Both of these schemes have impact on the burst size and the characteristics of the traffic injected into OBS core networks, thus affecting the edge-to-edge Quality of Service (QoS) provisioning.

In [25], the effect of different types of assembly mechanisms on Transmission Control Protocol (TCP) performance over OBS networks is studied. Three mechanisms are compared and evaluated, namely Fixed-Assembly-Period (FAP), Adaptive-Assembly-Period (AAP), and Min-BurstLength-Max-Assembly-Period (MBMAP). FAP is a timer-based assembly algorithm, in which a fixed assembly period is used by edge nodes to assemble IP packets with the same destination into one burst arriving at the fixed timeout period. AAP is also a timer-based algorithm with an adaptive assembly period adjusted according to the size of bursts recently sent. If the bursts recently sent are long, then AAP will increase the assembly period. Otherwise, the assembly period will be decreased. The rationale behind AAP is to make the burst assembly algorithm align with the TCP mechanism. If the bursts recently sent are long, it is very likely that TCP will send more traffic subsequently. On the other hand, if the bursts recently sent are short, it is possible that TCP is reducing its window size upon detecting a congestion. In this case, a short assembly period is favored. MBMAP is a mixed timer-based and threshold-based algorithm which has two control criteria for the burst assembly process: (i) the minimum burst length, and (ii) the maximum assembly period. Whenever the burst size exceeds the minimum burst length or the assembly timer expires, the burst assembly process stops. This algorithm avoids (i) sending bursts that are too small to the core networks, and (ii) delaying assembling IP packets for too long. Simulation results in [25] show that FAP achieves the best performance in terms of goodput improvement over the other two algorithms.

Based on this model, an adaptive timer-based assembly scheme which can support differentiated edge-to-edge latency requirements for multiple traffic types is proposed. In [26], an analytical model which derives the edge-to-edge delay for a timer-based assembly algorithm considering a Constant Bit Rate (CBR) traffic process is derived.

In [27], a threshold-based burst assembly scheme which can assign different thresholds to different traffic classes with differentiated QoS requirements is proposed. Simulation results in [27] show that there is an optimal threshold for each traffic class which minimizes the overall packet loss probability for a given network loading condition.

In [28], a composite burst assembly algorithm is proposed to assemble a burst from IP packets with different QoS requirements. The IP packets are placed from the head to the tail of a composite burst in order of decreasing priority. In case of burst contention in a core node, only the IP packets at the tail of the burst are dropped, thus providing priority and differentiated QoS to IP packets with higher priorities. Simulations results in [28] show that the proposed burst assembly algorithm can effectively support differentiated QoS to multiple traffic classes.

In [29, 30, 31], the effect of different burst assembly algorithms on the distribution of the output traffic from the assembler is studied. Both theoretical and simulation results show that the output traffic after the assembler approaches a Gaussian distribution under either Poisson or long range dependence input traffic. The variances of burst size and inter-arrival time decrease with increasing assembly window size and traffic load. The output traffic becomes smoother which helps to enhance the overall performance of OBS networks. But the long range dependence of the input traffic will remain unchanged after assembly. However, simulation results in [32] show that a simple timer-based burst assembly algorithm can reduce the self-similarity of the input traffic. It is also shown that the average assembly delay is bounded by the minimum burst size and timeout period of the assembler.

In [33, 34], various linear prediction based algorithms are proposed to set burst assembly thresholds according to predicted incoming traffic information. A header packet with the predicted incoming traffic information is sent to the core nodes to reserve bandwidth based on the predicted traffic information before the data burst is actually assembled. In this way, burst reservation and assembly can be done in parallel, thus reducing the edge-to-edge delay experienced by each burst.

1.2.3 Signaling mechanisms

In [26], a centralised version of OBS with two-way reservation for each burst, called wavelength routed OBS (WR-OBS), is proposed.

Before transmitting a burst, an ingress node sends a reservation message to a centralised server. For each reservation request, the server calculates the route from the ingress node to the egress node and reserves wavelengths at every link along the route for the burst. The burst is transmitted only after a successful acknowledgement message has been received from the server. It is claimed that WR-OBS improves network throughput and includes explicit QoS provisioning. However, the centralised nature of the scheme does not scale well and makes it unsuitable for large optical networks.

Unlike WR-OBS, there are three main one-way signaling mechanisms that differ mostly in the way wavelengths are reserved. In Just-in-Time (JIT) [16, 35], an output wavelength is reserved as soon as a header packet arrives at a node and is released only after a release message is received. This technique is simple to implement. However, it does not utilise the channels during the period between the arrival of a header and the arrival of the corresponding burst, which may be considerable. In Just-Enough-Time (JET) [15, 19], the time offset information is included in the header packet in addition to burst length information. This allows a core node to reserve a wavelength for a burst just before its actual arrival. Therefore, in the period between the header packet and burst arrival epochs, the channel can be used to transmit other bursts. This can lead to significant improvements in burst loss performance if the offset times in the network are large [19]. Thus, JET is probably the most popular OBS signaling scheme.

1.3 Quality of Service Support in OBS Networks

Due to the extreme popularity and success of the Internet, there is great diversity in current Internet applications with very different requirements of network performance or Quality of Service (QoS). Non-interactive and semi-interactive applications such as email and web browsing can cope with a wide range of QoS. On the other hand, highly interactive applications such as video conferencing and online gaming have very stringent operating requirements. In addition, not all users need the same level of QoS and may wish to pay the same price for it. Some companies that rely on the Internet for critical transactions may be willing to pay high premiums to ensure network reliability. In contrast, casual home users only need cheap Internet access and can tolerate a lower service level. The central point of this discussion is that some degree of controllability on the QoS provided to users is desirable, so that applications and users get the service level they need and at the same time, network service providers maximise their returns from the networks. We refer to such controllability of the QoS provided to users as QoS support.

In general, offering QoS support to end users, or end-to-end QoS provisioning, requires the participation of all network entities along the end-to-end paths. This is because the network performance perceived by an end user is the cumulative result of the service received by the user's packets at network entities along the end-to-end path. For example, consider an application that requires an end-to-end packet loss probability of no more than 1%. If the packet loss probability at just one single router on the path becomes larger than 1%, the required end-to-end QoS cannot be achieved. This requirement implies that OBS networks, which are a provisioning backbone network of the next-generation Internet, must have QoS support across ingress/egress pairs in order to realise edge-to-edge QoS support.

Closely related to QoS provisioning is the issue of network performance enhancement in general. To maximise profits, network operators would like to provide the required QoS levels with the least amount of network resources. Alternatively, they would like to provide QoS support for as many users as possible with a fixed amount of network resources. This applies for communication networks in general and OBS networks in particular. Therefore, if QoS provisioning algorithms are important to network users, QoS enhancement algorithms are equally important to network operators.

A solution used in wavelength-routed networks is to treat the optical connection between an ingress/egress pair as a virtual link.

The ingress and egress nodes then become adjacent nodes and QoS mechanisms developed for IP networks can be applied directly. This approach works well for wavelength-routed networks because wavelengths are reserved exclusively. Therefore, there is no data loss on the transmission path between an ingress node and an egress node, which makes the connection's characteristics resemble those of a real optical link. On the other hand, wavelengths in OBS networks are statistically shared among many connections between different source and destination pairs. Hence, there is a finite burst loss probability on the transmission path between an ingress node and an egress node, which renders this approach unusable for OBS networks.

Since OBS is similar to a datagram transport protocol like IP and there have been extensive QoS works for IP networks, it is desirable to adapt IP QoS solutions for use in OBS. However, there are unique features of OBS that must be considered in this process. In the following paragraphs, these differences will be discussed.

A primary difference between OBS and IP networks is that there is no or minimal buffering inside OBS networks. Therefore, an OBS node must schedule bursts as they arrive. This poses a great challenge in adapting IP-QoS solutions for OBS because most of the QoS differentiation algorithms in IP networks rely on the ability of routers to buffer and select specific packets to transmit next. It also makes it more difficult to accommodate high priority traffic classes with very low loss probability thresholds. For example, if two overlapping high priority bursts attempt to reserve a single output wavelength, one of them will be dropped. This is unlike the situation in IP networks where one of them can be delayed in a buffer while the other is being transmitted. Further, without buffering at core nodes, burst loss performance of a traffic class depends strongly on its burst characteristics. Bursts with long durations or short offset times are more likely to be dropped than others. Hence, it is difficult to have a consistent performance within one traffic class.

In summary, the diversity of Internet applications and users makes it desirable to have QoS support built into the Internet. In addition, from the network operators' point of view, QoS enhancement algorithms that maximise network performance are also important for economic reasons. As OBS is envisioned to be the optical transport architecture in the core of the Internet, it is imperative to develop QoS provisioning and enhancement algorithms for OBS networks. A possible approach is to modify QoS solutions designed for IP networks for use in OBS networks. However, there are unique features of OBS networks that must be taken into consideration. These features present new challenges and opportunities for QoS provisioning algorithms in OBS networks. Given that QoS support is an essential feature of any next generation network and OBS prospect as a next generation optical networking technology is increasingly promising, our book gives a comprehensive treatment on various QoS issues in OBS networks as described in next section, which is a timely update to the related book [18] on the general issues of OBS networks.

1.4 Overview

This book is organized into six chapters.

This chapter has given a brief introduction to the basic components and architecture of OBS networks, including burst assembly and signaling in OBS networks.

Chapter 2 will discuss the basic mechanisms to improve overall QoS in OBS networks. Some basic mechanisms discussed include burst scheduling, burst segmentation, burst rescheduling and ordered burst scheduling in OBS networks.

Chapter 3 will discuss relative QoS differentiation among multiple traffic classes in OBS networks. Various methods including offset time-based, burst segmentation, preemption based, header scheduling, priority based wavelength assignment, and proportional QoS differentiation methods will be discussed.

Chapter 4 will discuss absolute QoS provisioning in OBS networks. Various mechanisms such as offset-based mechanisms, virtual wavelength reservation, preemption-based mechanisms, burst early dropping, and wavelength grouping will be discussed.

Chapter 5 is devoted to the problem of edge-to-edge QoS provisioning in OBS networks. Some approaches including traffic engineering, admission control and reservation signaling and fairness provisioning will be discussed.

Chapter 6 will discuss some variants of OBS that have been reported in the literature and future research directions in OBS networks, including time-sliced OBS, wavelength routed OBS, and OBS Ring networks, Optical Burst Transport Networks (OBTN), and OBS testbeds.

References

1. C. Brackett, "Dense Wavelength Division Multiplexing Networks: Principles and Applications," *IEEE Journal on Selected Areas in Communications*, vol. 8, no. 6, pp. 948–964, 1990.
2. B. Mukherjee, *Optical WDM Networks*, Springer, 2006.
3. A. M. Glass et al., "Advances in Fiber Optics," *Bell Labs Technical Journal*, vol. 5, no. 1, pp. 168–187, 2000.
4. R. Ramaswami and K. N. Sivarajan, *Optical Networks: A Practical Perspective*, 2nd ed. Morgan Kaufmann Publishers, 2002.
5. A. R. Moral, P. Bonenfant, and M. Krishnaswamy, "The Optical Internet: Architectures and Protocols for the Global Infrastructure of Tomorrow," *IEEE Communications Magazine*, vol. 39, no. 7, pp. 152–159, 2001.
6. I. Chlamtac, A. Ganz, and G. Karmi, "Lightpath Communications: An Approach to High Bandwidth Optical WANs," *IEEE Transactions on Communications*, vol. 40, no. 7, pp. 1171–1182, 1992.
7. E. Modiano and P. J. Lin, "Traffic Grooming in WDM Networks," *IEEE Communications Magazine*, vol. 39, no. 7, pp. 124–129, 2001.
8. K. Zhu and B. Mukherjee, "Traffic grooming in an optical WDM mesh network," *IEEE Journal on Selected Areas in Communications*, vol. 20, no. 1, pp. 122–133, 2002.
9. L. Dittmann et al., "The European IST Project DAVID: A Viable Approach Toward Optical Packet Switching," *IEEE Journal on Selected Areas in Communications*, vol. 21, no. 7, pp. 1026–1040, 2003.
10. T. S. El-Bawab and J.-D. Shin, "Optical Packet Switching in Core Networks: Between Vision and Reality," *IEEE Communications Magazine*, vol. 40, no. 9, pp. 60–65, 2002.
11. D. K. Hunter and I. Andonovic, "Approaches to Optical Internet Packet Switching," *IEEE Communications Magazine*, vol. 38, no. 9, pp. 116–122, 2000.
12. M. J. O' Mahony, D. Simeonidou, D. K. Hunter, and A. Tzanakaki, "The Application of Optical Packet Switching in Future Communication networks," *IEEE Communications Magazine*, vol. 39, no. 3, pp. 128–135, 2001.
13. S. Yao, B. Mukherjee, S.J.B. Yoo, and S. Dixit, "A unified study of contention-resolution schemes in optical packet-switched networks," *Journal of Lightwave Technology*, vol. 21, no. 3, pp. 672–683, 2003.

14. Y. Chen, C. Qiao, and X. Yu, "Optical Burst Switching: A New Area in Optical Networking Research," *IEEE Network*, vol. 18, no. 3, pp. 16–23, 2004.
15. C. Qiao and M. Yoo, "Optical Burst Switching - A New Paradigm for An Optical Internet," *Journal of High Speed Network*, vol. 8, no. 1, pp. 69–84, 1999.
16. J. Y. Wei and R. I. McFarland Jr., "Just-in-Time Signaling for WDM Optical Burst Switching Networks," *IEEE/OSA Journal of Lightwave Technology*, vol. 18, no. 12, pp. 2019–2037, 2000.
17. Y. Huang, D. Datta, J. P. Heritage, Y Kim, B. Mukherjee, "A Novel OBS Node Architecture using Waveband-Selective Switching for Reduced Component Cost and Improved Performance," in *Proc. IEEE LEOS*, 2004, pp. 426–427.
18. J.P. Jue and V.M Vokkarane, *Optical Burst Switched Networks*, Springer, Optical Networks Series, 2005.
19. Y. Xiong, M. Vandenhoute, and H. C. Cankaya, "Control Architecture in Optical Burst-Switched WDM Networks," *IEEE Journal on Selected Areas in Communications*, vol. 18, no. 10, pp. 1838–1851, 2000.
20. Traffic Control and Congestion Control in B-ISDN, Recommendation I.371, ITU-T, 1995.
21. E. A. Varvarigos and V. Sharma, "The Ready-to-Go Virtual Circuit Protocol: A Loss-Free Protocol for Multigigabit Networks using FIFO Buffers," *IEEE/ACM Transactions on Networking*, vol. 5, no. 5, pp. 705–718, 1997.
22. I. Widjaja, "Performance Analysis of Burst Admission-Control Protocols," *IEE Proceedings - Communications*, vol. 142, no. 1, pp. 7–14, 1995.
23. E. Rosen, A. Viswanathan, and R. Callon, "Multiprotocol Label Switching Architecture," RFC 3031, 2001.
24. C. Qiao, "Labeled Optical Burst Switching for IP-over-WDM Integration," *IEEE Communications Magazine*, vol. 38, no. 9, pp. 104–114, 2000.
25. X. Cao, J. Li, Y. Chen, and C. Qiao, "TCP/IP Packets Assembly over Optical Burst Switching Network," in *Proc. IEEE Globecom*, 2002, pp. 2808–2812.
26. M. Duser and P. Bayvel, "Analysis of a Dynamically Wavelength-Routed Optical Burst Switched Network Architecture," *IEEE/OSA Journal of Lightwave Technology*, vol. 20, no. 4, pp. 574–585, 2002.
27. V. Vokkarane, K. Haridoss, and J.P. Jue, "Threshold-Based Burst Assembly Policies for QoS Support in Optical Burst-Switched Networks," in *Proc. Opticomm*, 2002, pp. 125–136.
28. V. Vokkarane, Q. Zhang, J.P. Jue, and B. Chen, "Generalized Burst Assembly and Scheduling Techniques for QoS Support to Optical Burst-Switched Networks," in *Proc. IEEE Globecom*, 2002, pp. 2747–2751.
29. X. Yu, Y. Chen, and C. Qiao, "Study of Traffic Statistics of Assembled Burst Traffic in Optical Burst Switched Networks," in *Proc. Opticomm*, 2002, pp. 149–159.
30. X. Yu, Y. Chen, and C. Qiao, "Performance Evaluation of Optical Burst Switching with Assembled Burst Traffic Input," in *Proc. IEEE Globecom*, 2002, pp. 2318–2322.
31. M. Izal and J. Aracil, "On the Influence of Self-Similarity on Optical Burst Switching Traffic," in *Proc. IEEE Globecom*, 2002, pp. 2308–2312.
32. A. Ge, F. Callegati, and L.S. Tamil, "On Optical Burst Switching and Self-Similar Traffic," *IEEE Communications Letters*, vol. 4, no. 3, pp. 98–100, 2000.

33. D. Morato, J. Aracil, L.A. Diez, M. Izal, and E. Magana, "On Linear Prediction of Internet Traffic for Packet and Burst Switching Networks," in *Proc. Tenth International Conference on Computer Communications and Networks*, pp. 138 –143, 2001.
34. J. Liu, N. Ansari, and T. J. Ott, "FRR for Latency Reduction and QoS Provisioning," *IEEE Journal on Selected Area in Communications*, vol. 21, no 7, pp. 1210–1219, 2003.
35. A. Zalesky, H. L. Vu, Z. Rosberg, E. W. M. Wong, and M. Zukerman, "Modelling and Performance Evaluation of Optical Burst Switched Networks with Deflection Routing and Wavelength Reservation," in *Proc. IEEE Infocom*, 2004, pp. 1864–1871.

2

NODE-BASED QOS IMPROVEMENT MECHANISMS

Quality of Service (QoS) mechanisms in a network can be broadly divided into two categories: QoS improvement and QoS provisioning mechanisms. A QoS improvement mechanism can be defined as any mechanism that improves the general performance of the network. Although less obvious than QoS provisioning mechanisms, QoS improvement mechanisms are very important in enabling the network to provide satisfactory service to end users. They allow the network to accommodate more users and reduce the cost of data transmission.

This chapter focuses on QoS improvement mechanisms located at a node in an OBS network. A survey of the current state-of-the-art, including optical buffering, deflection routing, burst segmentation, wavelength conversion and channel scheduling is presented. Then two channel scheduling algorithms that take advantage of the offset times, a unique feature of OBS, to give good performance are presented in detail.

2.1 Contention Resolution Approaches

Since a wavelength channel may be shared by many connections in OBS networks, there exists the possibility that bursts may contend with one another at intermediate nodes. Contention occurs when multiple bursts from different input ports are destined for the same output port simultaneously. The general solution to burst contention is to move all but one burst "out of the way". An OBS node has three possible dimensions to move contending bursts, namely, time, space and wavelength. The corresponding contention resolution approaches are optical buffering, deflection routing and wavelength conversion, respectively. In addition, there is another approach unique to OBS called burst segmentation.

2.1.1 Optical buffering

Typically, contention resolution in traditional electronic packet switching networks is implemented by storing excess packets in Random Access Memory (RAM) buffers. However, RAM-like optical buffers are not yet available. Currently, optical buffers are constructed from Fibre Delay Lines (FDLs) [1, 2, 3]. An FDL is simply a length of fibre and hence offers a fixed delay. Once a packet/burst has entered it, it must emerge after a fixed length of time later. It is impossible to either remove the packet/burst from the FDL earlier or hold it in the FDL longer. The fundamental difficulty facing the designer of an optical packet/burst switch is to implement variable-length buffers from these fixed-length FDLs.

Current optical buffers may be categorised in different ways. They can be classified as either single-stage, i.e., having only one block of parallel delay lines, or multi-stage, which have several blocks of delay lines cascaded together. Single-stage optical buffers are easier to control, but multi-stage implementations may lead to more savings on the amount of hardware used. Optical buffers can also be classified as having feed-forward or feedback configurations. In a feed-forward configuration, delay lines connect the output of a switching stage to the input of the next switching stage. In a feedback configuration, delay lines connect the output of a switching stage back to the input of the same stage. Long holding time and certain degrees of variable delays can be easily implemented with a feedback configuration by varying the number of loops a packet/burst undergoes. However, each loop causes some loss in signal power. Therefore, a packet/burst cannot be stored indefinitely in a feedback architecture. In a feed-forward configuration, delay lines with different lengths must be used to achieve variable delays. This architecture attenuates all signals almost equally because every packet/burst passes through the same number of switches. Hybrid combinations of feedforward and feedback architectures are also possible [4].

Based on the position of buffers, packet switches fall into one of three major categories: input buffering, output buffering and shared buffering. In input-buffered switches, a set of buffers is assigned for each input port. This configuration has poor performance due to the head-of-line blocking problem. Consequently, it is never proposed for purely optical implementation. In output-buffered switches, a set of buffers is assigned to each output port. Most optical switches emulate output buffering since the delay in each output optical buffer can be determined before the packet/burst enters it. Shared buffering is similar to output buffering except that all output ports share a common pool of buffers.

Due to their hardware-saving characteristics, multi-stage and/or shared-buffered architectures are predominant in optical switch proposals. Figure 2.1 shows two single-stage, shared-buffered switch architectures [5] with feedforward and feedback configurations where N and B are the number of input ports and the number of FDLs, respectively. They both contain an FDL pool that is shared among all output ports. In the feedforward configuration, packets/bursts may be delayed only once, whereas the feedback configuration allows them to be delayed multiple times. Since the FDLs are optical fiber themselves, it is possible for them to hold multiple packets/bursts of different wavelengths simultaneously [6]. However, this comes at the expense of increased complexity in scheduling algorithms. Compared to single-stage buffer architectures, multi-stage counterparts [7, 8, 9] are much more complex. They contain several primitive switching elements connected together by FDLs, usually in a feedforward configuration. Multi-stage buffers can achieve buffer depth of several thousands.

Recently, optical buffers based on slow-light delay lines have received considerable interest [10]. In slow-light delay lines, light is slowed down using a variety of techniques such as electromagnetically induced transparency (EIT), population oscillations (POs) and microresonator-based photonic-crystal (PC) filter. In principle, these techniques can make the group velocity approach zero.

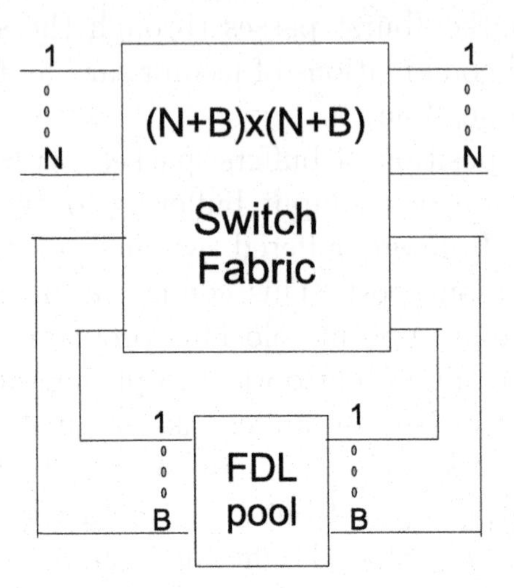

(a) Feedback shared-buffered architecture

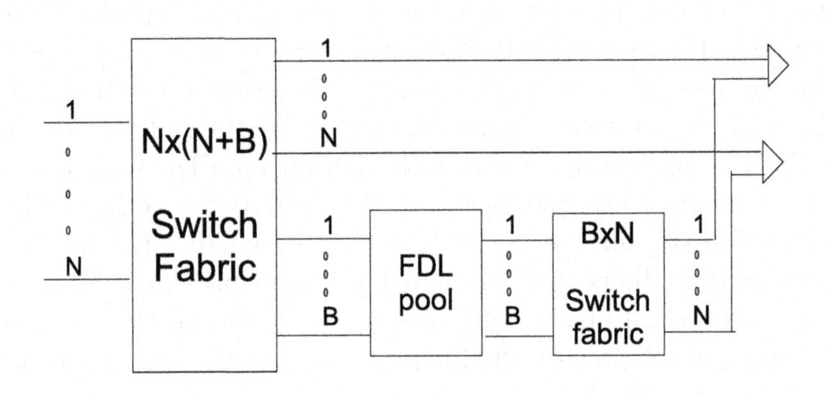

(b) Feedforward shared-buffered architecture

Fig. 2.1. Single-stage optical buffer architectures. ©[2006] IEEE.

However, very slow group velocity always comes at the cost of very low bandwidth or throughput. Therefore, slow-light delay lines are still not practical in optical switches that have to handle very high data rates.

In summary, despite the considerable research efforts on FDL-based optical buffers, there remain some hurdles that limit their effectiveness. Firstly, by their nature, they can only offer discrete delays. The use of recirculating delay lines can give finer delay granularity but it also degrades optical signal quality. Secondly, the size of FDL buffers is severely limited not only by signal quality concerns but also by physical space limitations. A delay of 1 ms requires over 200 km of fibre. Due to the size limitations of buffers, optical buffering alone as a means of contention resolution may not be effective under high load or bursty traffic conditions.

2.1.2 Deflection routing

Deflection routing is a contention resolution approach ideally suited for photonic networks that have little buffering capacity at each node. If no buffer is present, deflection routing is also known as hot-potato routing. In this approach, if the intended output port is busy, a burst/packet is routed (or deflected) to another output port instead of being dropped. The next node that receives the deflected burst/packet will try to route it towards the destination. The performance of slotted deflection routing has been extensively evaluated for regular topologies such as ShuffleNet, hypercube and Manhattan Street Network [11, 12, 13]. It is found that deflection routing generally performs poorly compared to store-and-forward routing unless the topology in use is very well connected. Nevertheless, its performance can be significantly improved with a small amount of buffers. Between slotted and unslotted networks, deflection routing usually performs better in the former since the networks can make use of the synchronous arrival of the packets to a router to minimise locally the number of deflections. Nevertheless, such deflection minimisation can also be done to some extent in unslotted networks using heuristics [14]. This brings the performance of deflection routing in unslotted networks close to that in slotted networks.

For an arbitrary topology, the choice of which output links to use for deflected bursts/packets is critical to the performance of the network. The existing deflection routing protocols can be divided into three categories: fixed alternate routing, dynamic traffic aware and random routing. Fixed alternate routing is the most popular approach. In this method, the alternate path is either defined on a hop-by-hop basis [15] or by storing at each node both the complete primary path and the complete alternate path from itself to every possible destination node in the network [16]. Fixed alternate routing can yield good performance on small topologies. However, selecting a good alternate path becomes difficult on large topologies due to the tight coupling between subsequent burst loss probabilities, traffic matrices and network topology. Traffic aware deflection routing takes into consideration the transient traffic condition in selecting the output links for deflected bursts/packets [17, 18]. It becomes similar to load balancing, which we will address in Chapter 5. Random deflection routing [19] appears to strike the right balance between simplicity of implementation, robustness and performance. In this approach, bursts/packets carry in their header a priority field. Every time a burst/packet is deflected, its priority is decreased by one. Normal bursts/packets on their primary paths can preempt those low priority ones. Thus, the worst-case burst/packet loss probability of this method is upper-bounded by that in standard networks.

To apply deflection routing to OBS networks, the problem of insufficient offset time must be overcome. This problem is caused by a burst traversing more hops than originally intended as a result of being deflected. Since the offset time between the burst and its header decreases after each hop, the burst may overtake the header packet. Various solutions have been proposed [16], such as setting extra offset time or delaying bursts at some nodes on the path. It is found that delaying a burst at the next hop after it is deflected is the most promising option.

Deflection routing may be regarded as "emergency" or *unplanned* multipath routing. It might cause deflected bursts to follow a longer path than other bursts in the same flow. This leads to various problems such as increased delay, degradation of signal quality, increased network resource consumption and out-of-order burst arrivals. A better method to reduce congestion and burst loss is probably *planned* multipath routing, or load-balancing. The topic of load balancing will be discussed in chapter 5.

2.1.3 Burst segmentation

Burst segmentation [20, 21] is a contention resolution approach unique to OBS networks. It takes advantage of the fact that a burst is composed of multiple IP packets, or segments. Therefore, in a contention between two overlapping bursts, only the overlapping segments of a burst need to be dropped instead of the entire burst. Network throughput is improved as a result. Two currently proposed variants of burst segmentation are shown in Figure 2.2. In the head-dropping variant [20], the overlapping segments of the later arriving burst, or the head segments, are dropped. On the other hand, the tail-dropping variant [21] drops the overlapping segments of the preceding burst, or the tail segments. A number of strategies to combine burst segmentation with deflection routing have also been discussed. Comparing the two variants, the tail-dropping approach results in a better chance of in-sequence delivery of packets at the destination. Burst segmentation is later integrated with void-filling scheduling algorithms in [22, 23]. A performance analysis of burst segmentation is presented in [24].

2.1.4 Wavelength conversion

Wavelength conversion is the process of converting the wavelength of an incoming signal to another wavelength for transmission on an outgoing channel. In WDM, each fibre has several wavelengths, each of which functions as a separate transmission channel. When contention for the same output wavelength happens between some bursts, the node equipped with wavelength converters can convert all except one burst to other free wavelengths. Wavelength conversion enables an output wavelength to be used by bursts from several input wavelengths, thereby increasing the degree of statistical multiplexing and the burst loss performance. As the number of wavelengths that can be coupled into a fibre continues to grow, this approach becomes increasingly attractive. For example, with 32 wavelengths per link, the burst loss probability at a loading of 0.8 is about 4×10^{-2}. With 256 wavelengths per link, the burst loss probability drops to less than 10^{-4}.

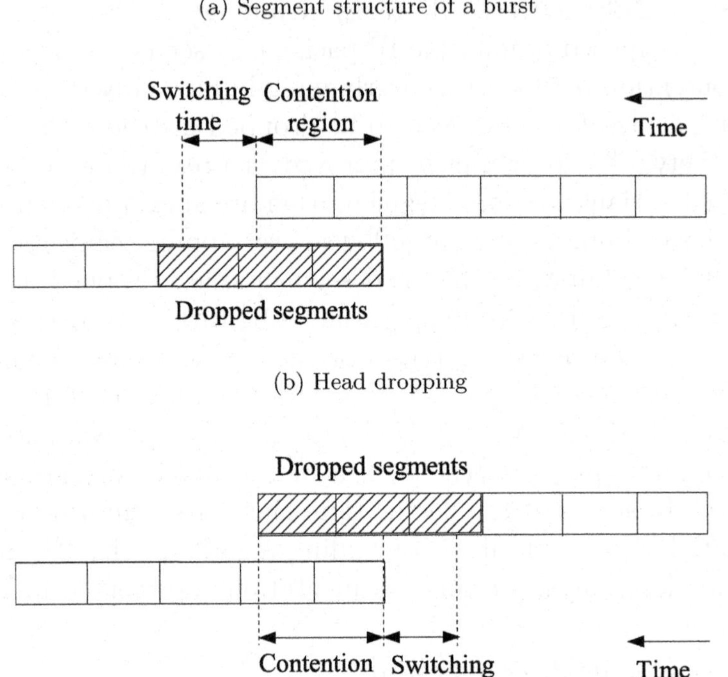

(a) Segment structure of a burst

(b) Head dropping

(c) Tail dropping

Fig. 2.2. Burst segmentation approaches

Although optical wavelength conversion has been demonstrated in the laboratory environment, the technology remains expensive and immature. Therefore, to be cost-effective, an optical network may be designed with some limitations on its wavelength conversion capability. Following are the different categories of wavelength conversion:

- *Full conversion:* Any incoming wavelength can be converted to any outgoing wavelength at every core node in the network. This is assumed by most current OPS and OBS proposals. It is the best performing and also the most expensive type of wavelength conversion.

- *Sharing of converters at a node:* Converter sharing [25, 26, 27] is proposed for OPS/OBS networks. It allows savings on the number of converters needed. However, the drawbacks are the enlargement of the switching matrix and additional attenuation of the optical signal.
- *Sparse location of converters in the network:* Only some nodes in the network are equipped with wavelength converters. Although this category is well-studied for wavelength-routed networks, it has not been widely considered for OPS and OBS networks due to the poor loss performance at nodes without wavelength conversion capability.
- *Limited-range conversion:* An incoming wavelength can only be converted to some of the outgoing wavelengths. Various types of limited-range converters for OPS networks have been examined [28, 29, 30]. It is shown that nodes with limited-range wavelength converters can achieve loss performance close to those with full conversion capability.

2.2 Traditional Channel Scheduling Algorithms

Since the large number of wavelengths per link in WDM offers excellent statistical multiplexing performance, wavelength conversion is the primary contention resolution approach in OBS. In this approach, every OBS core node is assumed to have full wavelength conversion capability. When a header packet arrives at a node, the node invokes a channel scheduling algorithm to determine an appropriate outgoing channel to assign to the burst. Channel scheduling plays a crucial role in improving the burst loss performance of an OBS switch. A good scheduling algorithm can achieve several orders of magnitude performance improvement over a first-fit algorithm. Because of its importance, channel scheduling in OBS has been the subject of intense research in the last few years.

In the JET OBS architecture, each burst occupies a fixed time interval, which is characterised by the start time and the end time carried in the header packet. Therefore, channel scheduling can be

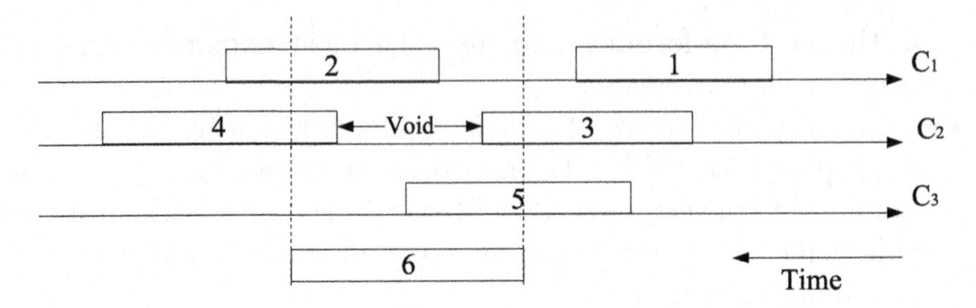

Fig. 2.3. Illustration of the channel fragmentation problem

regarded as a packing problem wherein the primary objective is to pack as many incoming bursts onto the outgoing channels as possible. This problem is complicated by the fact that the order of the header packet arrivals is not the same as the arrival order of the bursts themselves. Thus, bursts with long offset times are able to reserve a channel before those with shorter offset times. Their reservations fragment a channel's free time and produce gaps or *voids* among them that degrade the schedulability of bursts with shorter offset times. This is illustrated in Figure 2.3 where the numbers inside the bursts indicate their header arrival order. Although all six bursts can theoretically be accommodated, burst 6 cannot be scheduled because of the channel fragmentation caused by the other bursts. Many channel scheduling algorithms have been proposed to deal with this problem. In this section, a survey of some traditional channel scheduling algorithms is given.

2.2.1 Non-void filling algorithm

Non-void filling algorithm is the simplest type of channel scheduling algorithms. It is named Horizon [31] and Latest Available Unscheduled Channel (LAUC) [32] by two independent research groups. In order to maximise processing speed, it does not utilise voids caused by previously scheduled bursts to schedule new bursts. Instead, it only keeps track of the *unscheduled* time, which is the end time of the last scheduled burst, for each channel. When a header arrives, it assigns to the burst the channel with the unscheduled time being closest but not exceeding its start time. The idea is to minimise the void produced before it. This is illus-

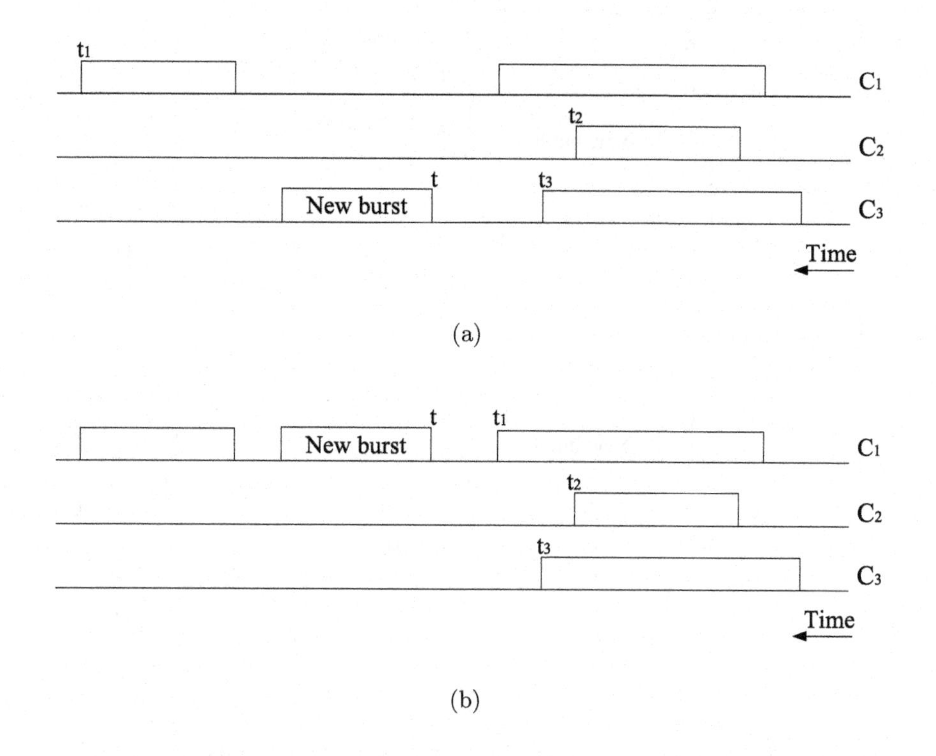

Fig. 2.4. Examples of channel assignment using: (a) non-void filling algorithm (Horizon or LAUC), and (b) void filling algorithm (LAUC-VF)

trated in Figure 2.4(a) where t_1, t_2 and t_3 are the unscheduled times. Channel C_3 is selected to schedule the new burst because $t - t_3 < t - t_2$. By storing the unscheduled times in a binary search tree, the two algorithms can be executed in $O(logW)$ time, where W is the number of wavelengths per link.

2.2.2 Algorithms with void filling

Void-filling algorithms [32, 33] utilise voids to schedule new bursts to improve burst loss performance. They keep track of every void on the outgoing channels and check all of them as well as unscheduled channels when an incoming burst needs to be scheduled. Latest Available Unused Channel with Void Filling (LAUC-VF) [32] is perhaps the most popular OBS channel scheduling algorithm to date. When an incoming burst needs to be scheduled, LAUC-VF

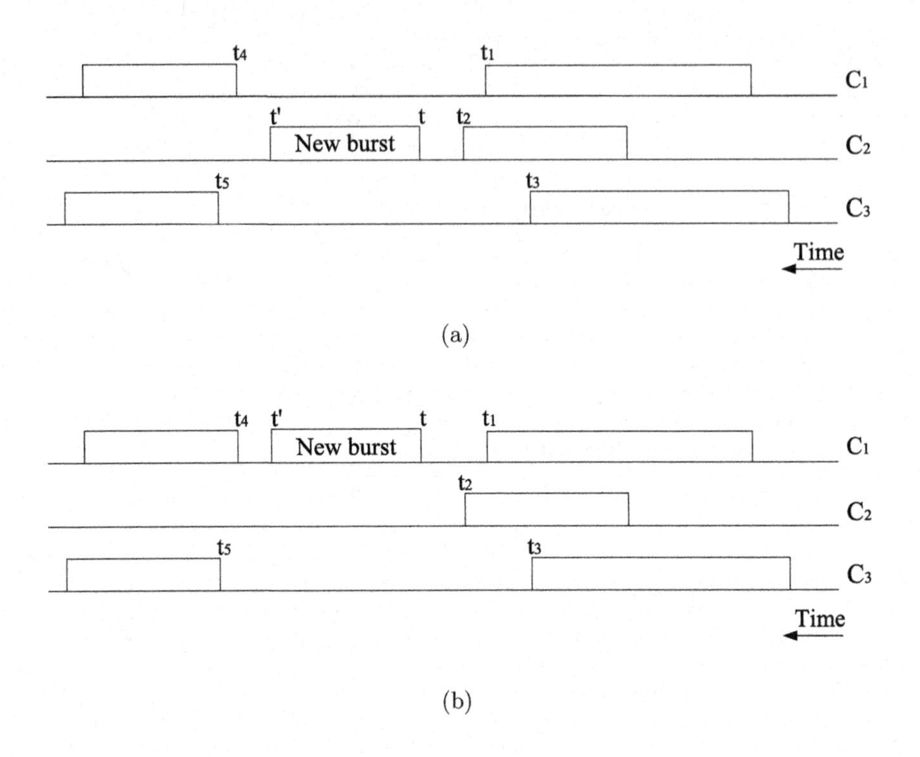

Fig. 2.5. Comparison of LAUC-VF and its variants: (a) LAUC-VF, and (b) LAUC-VF variants

calculates the *unused* time of each available channel, which is the end time of the burst preceding the incoming one. The channel with the unused time closest to the start of the incoming burst is selected. This is illustrated in Figure 2.4(b) where t_1, t_2 and t_3 are the unused times. Channel C_1 is selected to schedule the new burst because t_1 is closest to t. Since the unused times for each burst are different, LAUC-VF has to recalculate all of them for each new burst. Using a binary search tree structure, each unused time calculation takes $O(logN_b)$, where N_b is the average number of scheduled bursts per channel. Thus, LAUC-VF takes $O(WlogN_b)$ to execute.

A drawback of the basic LAUC-VF algorithm is that it may select an unscheduled channel to schedule a new burst even though suitable voids are available because it bases its decision only on the unused times. This problem is illustrated in Figure 2.5(a) where

LAUC-VF selects channel 2 for the new burst because $t - t_2$ is smaller than both $t - t_1$ and $t - t_3$. This creates more voids and degrades performance. An LAUC-VF variant [34] that solves the problem is to give priority to channels with voids, thus minimising the number of voids generated. A further refinement of LAUC-VF [35] is to compare all the voids that would be generated and choose the channel that will give minimum voids. In the illustration in Figure 2.5(b), the first variant will schedule the new burst on either channel 1 or channel 3 while the second variant will only choose channel 1 since $t-t_1 < t-t_3$ and $t_4-t' < t_5-t'$. Parallel processing and associative memory can be used to implement LAUC-VF and its variants to reduce the time complexity [36].

Another implementation of LAUC-VF [37] is called Minimum Starting Void (Min-SV). Min-SV has the same scheduling criteria as LAUC-VF. However, it uses an augmented balanced binary search tree as the data structure to store the scheduled bursts. This enables Min-SV to achieve processing time as low as that of Horizon without requiring special hardware or parallel processing.

2.3 Burst-Ordered Channel Scheduling Approach

A common characteristic of the traditional channel scheduling algorithms is that they schedule incoming bursts in the order of their header packet arrivals. Therefore, the bursts with long offset times are able to reserve a channel before those with shorter offset times. Their reservations fragment a channel's free time and produce voids among them that degrade the schedulability of bursts with shorter offset times. To address the root of this problem, it is required that a node schedules incoming bursts in the order of their actual arrivals. In that case, the voids will have no negative effect on burst scheduling because by the time a void is generated, all the bursts in that region have already been scheduled.

The most popular method to implement the burst-ordered scheduling approach is to delay the processing of headers that arrive early (bursts with long offset times) in order to collect information from the headers that arrive later (bursts with short offset times). In the batch scheduling variant [38, 39], a node collects multiple burst headers and makes scheduling decisions for all of them at one go. In the delayed scheduling variant [40, 41], the node delays burst headers for a certain period and sorts them according to their burst arrivals. However, in both variants, the inherent conflict between the need to delay headers and the need to forward headers early to give downstream nodes sufficient processing times prevents the burst-ordered scheduling concept from being fully realised.

Dual-header OBS (DOBS) [42] is another method to implement the burst-ordered scheduling concept. In this method, information about each burst is carried in two separate headers: the *service request packet* (SRP) header and the *resources allocated packet* (RAP) header. The SRP header originates from the ingress node and carries service requirement information such as offset time and burst length, which allows intermediate nodes to calculate the time interval that the incoming burst will occupy. Upon receiving an SRP header, a node records the information contained within the header and passes it onto the next downstream node. Later, just before the corresponding data burst arrives, the node makes a scheduling decision for the burst and sends the scheduled channel information to the next node in an RAP header. This method resolves the conflict in the batch scheduling and delayed scheduling methods and fully realises the burst-ordered scheduling concept. However, it has the problem of phantom reservations since a SRP header is sent to the next node before the scheduling decision is made. Therefore, it is possible that the burst is dropped at that node while the SRP header continues to make reservations at downstream nodes. Those phantom reservations caused by bursts dropped at an upstream node lock up resources at downstream nodes, which could be used for other bursts.

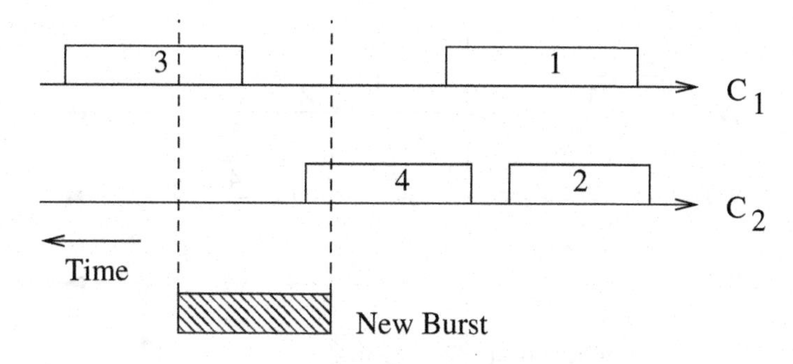

(a) LAUC-VF fails to schedule the new burst

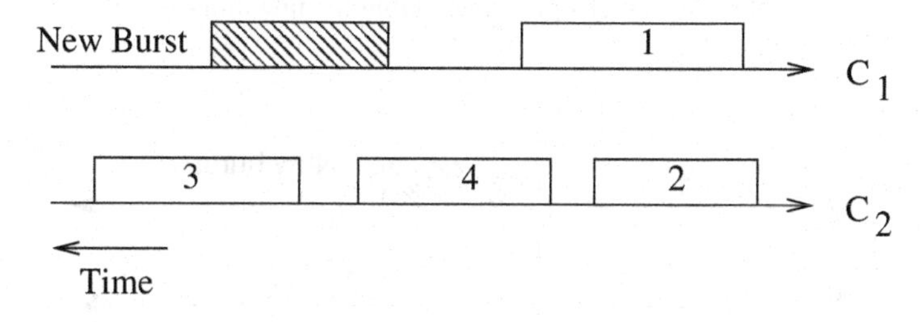

(b) Burst 3 is rescheduled to accommodate the new burst

Fig. 2.6. Illustration of the benefits of burst rescheduling

2.4 Burst Rescheduling

Burst rescheduling[1] [43, 44] attempts to approximate the burst-ordered scheduling concept. It helps to improve the burst loss performance and at the same time achieve low computational complexity. The key idea of burst rescheduling is to reschedule an existing scheduled burst to another wavelength to accommodate an incoming burst. This is possible as requests arrive dynamically and a header packet reserves wavelengths well before the arrival of its corresponding data burst. The benefit of burst rescheduling is illustrated in Figure 2.6. In this scenario, the new burst, which cannot be scheduled by LAUC-VF, can be scheduled after

[1] Reprinted from (S. K. Tan, G. Mohan, and K. C. Chua, "Algorithms for Burst Rescheduling in WDM Optical Burst Switching Networks," *Computer Networks*, vol. 41, no. 1, pp. 41–55), ©[2003], with permission from Elsevier.

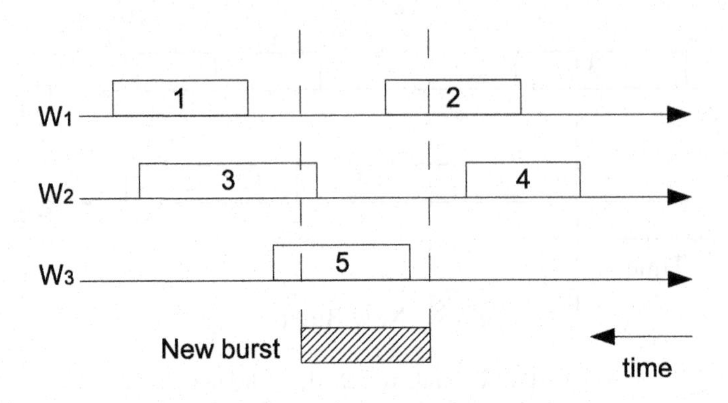

(a) No wavelength is available for new burst.

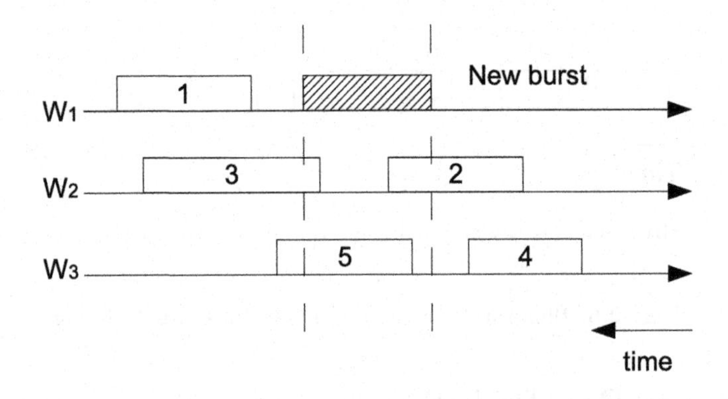

(b) Multi-level rescheduling to accommodate new burst.

Fig. 2.7. Illustration of multi-level rescheduling.

burst 3 is moved to another free wavelength. It is to be noted that rescheduling does not affect any ongoing traffic. Rescheduling a burst on a link requires changes in the control setting in both the end nodes of the link. Therefore, whenever rescheduling is successful, a special "NOTIFY" packet is sent to the next node to notify it about the changes, for e.g., wavelength for that burst so that the receiving node will do the necessary settings.

Rescheduling algorithms can be developed based on two approaches. These are *single-level rescheduling* and *multi-level rescheduling*. Single-level rescheduling involves the rescheduling of only one burst to another available wavelength to accommodate a new burst. The example illustrated in Figure 2.6(b) falls under this category. In multi-level burst rescheduling, several bursts are rescheduled one by one in sequence to other available wavelengths in order to accommodate a new burst. As shown in Figure 2.7(a), no wavelength is available for the new burst if single-level rescheduling is used. Multi-level rescheduling can reschedule burst 4 from W_2 to W_3 followed by rescheduling burst 2 from W_1 to W_2 to free wavelength W_1 to accommodate the new burst, as shown in Figure 2.7(b). Multi-level rescheduling is expected to provide better performance than single-level rescheduling. However, from the computational complexity point of view, multi-level rescheduling is more complex than single-level rescheduling. This is because a multi-level rescheduling algorithm needs to determine an appropriate order (among several possibilities) in which different bursts are to be rescheduled in sequence. Since the objective is to achieve low computational complexity, single-level rescheduling is generally preferred.

2.4.1 Burst rescheduling algorithms

On-Demand Burst Rescheduling (ODBR) algorithm

As the name suggests On-Demand Burst Rescheduling (ODBR) algorithm considers rescheduling of an existing burst only when a burst fails to be scheduled to any of the wavelengths. The algorithm works in two phases. When a new burst arrives, phase 1 is executed to select a suitable free wavelength using LAUC. If no wavelength is available, phase 2 is called to check if any of the existing bursts can be moved to a new wavelength to enable scheduling of the new burst. The algorithm examines the wavelengths one by one. For a given wavelength, it checks if the last burst can be moved to any other wavelength and determines the void created. After examining all the wavelengths, it chooses the one which possibly creates the smallest void after migration.

The pseudo-code of the algorithm is given in Table 2.1. The new burst is assumed to arrive at time t. The latest available time of wavelength channel W_i is denoted by t_i.

Table 2.1. ODBR algorithm

Phase 1
Use LAUC to find a suitable wavelength for the incoming burst. If no such wavelength is found, call phase 2. Otherwise, exit.
Phase 2
Step 1: For every wavelength W_i and out-wavelength V_i, determine if V_i is valid. Out-wavelength V_i is said to be valid if the last burst on W_i can be moved to V_i and the new burst can be scheduled to W_i.
Step 2: If no valid out-wavelength V_i exists, the new burst is dropped. Otherwise, choose wavelength W_p that has a valid out-wavelength V_p and is the latest available wavelength after rescheduling among all the valid out-wavelengths.
Step 3: Reschedule last burst on W_p to V_p. Assign new burst to W_p.
Step 4: Send a special NOTIFY header packet to notify the next node about the change in wavelength of the rescheduled burst.

A simple example shown in Figure 2.8(a) and (b) helps to illustrate how ODBR works. Phase 1 fails to assign any wavelength to the new burst as shown in Figure 2.8(a) and phase 2 will therefore be invoked. Wavelength W_1 and W_3 both have a valid out-wavelength W_2. Therefore, rescheduling of the last burst from W_1 or W_3 to W_2 would make a wavelength available for scheduling the new burst. In order to optimize the performance, ODBR chooses the best wavelength which has the latest available time. In this case, the void formed by the new burst on W_3 by rescheduling the burst from W_3 to W_2 is the smallest compared to the void formed at W_1 by rescheduling the burst from W_1 to W_2. Therefore, W_3 is the best wavelength. The last burst on W_3 is rescheduled to W_2 and the new burst can be scheduled to W_3 as shown in Figure 2.8(b).

The complexity of ODBR is as follows

- *Phase 1 - Scheduling:* ODBR examines the information of one burst on each wavelength. Phase 1 runs in $O(W)$ time in the worst case.
- *Phase 2 - Rescheduling:* Phase 2 has the worst case complexity of $O(W^2)$ time since it examines the last burst on each wavelength for rescheduling to one of the other wavelengths.

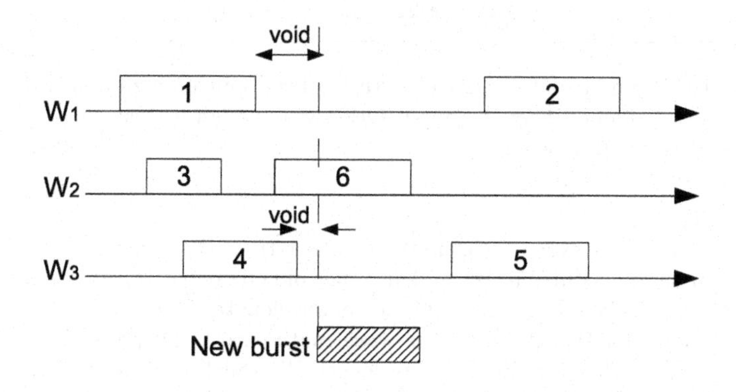

(a) The new burst cannot be scheduled.

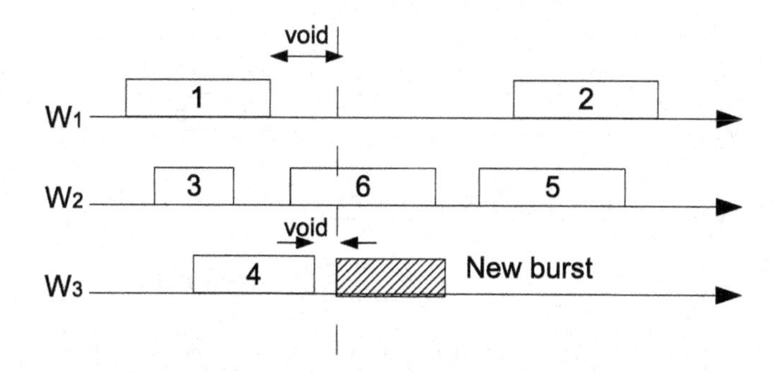

(b) The last burst on W_3 is moved to W_2 to accommodate the new burst on W_3.

Fig. 2.8. Illustration of multi-level rescheduling.

Since the complexity of LAUC-VF is $O(KW)$ with K being the average number of scheduled bursts per wavelength, the complexity of ODBR will be more than LAUC-VF if $W > K$. However, since ODBR is called only when a burst is dropped (usually less than 10%), the overall processing complexity remains better than LAUC-VF. It therefore has the advantage of low complexity similar to LAUC.

Aggressive Burst Rescheduling (ABR) algorithm

As shown in Table 2.2, the ABR algorithm also has two phases. However, it is different from the ODBR algorithm in that phase 2

Table 2.2. ABR algorithm

Phase 1
Use LAUC to find a suitable wavelength W_p for the incoming burst. If no such wavelength is found, drop the burst. Otherwise, assign wavelength W_p to the new burst and call phase 2.
Phase 2
Step 1: For every wavelength W_i other than W_p determine if the last burst can be rescheduled to W_p and also the void created at W_p after rescheduling. If such rescheduling is possible for a wavelength, it is said to be a valid in-wavelength for W_p. If no valid in-wavelength exists, exit phase 2.
Step 2: Choose a valid in-wavelength W_j which has the smallest void.
Step 3: Reschedule the last burst from W_j to W_p.
Step 4: Send a special NOTIFY header packet to notify the next node about the change of wavelength of the rescheduled burst.

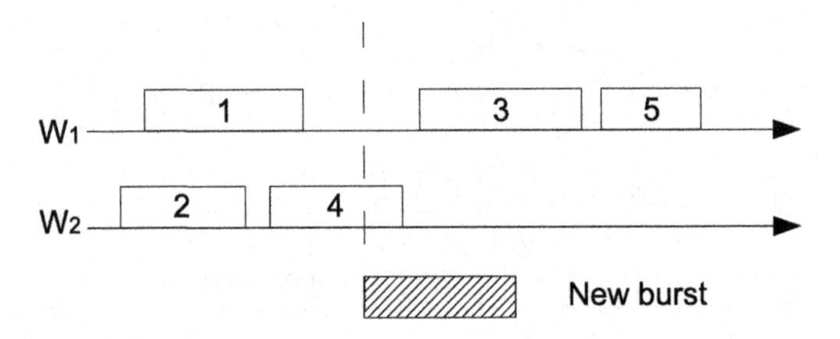

Fig. 2.9. LAUC, ODBR and LAUC-VF fail to schedule new burst 6.

is not invoked when phase 1 fails but when phase 1 is successful. This algorithm is intended to prevent future data burst dropping by invoking rescheduling every time a burst has been scheduled successfully. In ABR, upon successful scheduling of a burst at W_p in phase 1, rescheduling of one latest burst from some other wavelength W_i to W_p takes place in phase 2 if such a burst exists. Rescheduling is governed by the rule that the void formed when the burst is rescheduled from W_i to W_p is minimum among all possible wavelengths. By doing so, the probability of dropping data bursts that arrive later could be decreased.

Examples as shown in Figure 2.9 and Figure 2.10 are used to illustrate this algorithm. Figure 2.9 shows two wavelengths and bursts that are being considered. Burst 1, 2, 3, 4, and 5 arrive at a node one by one in that order and are scheduled to W_1 and W_2 at phase 1. When burst 6 arrives, it cannot be scheduled to

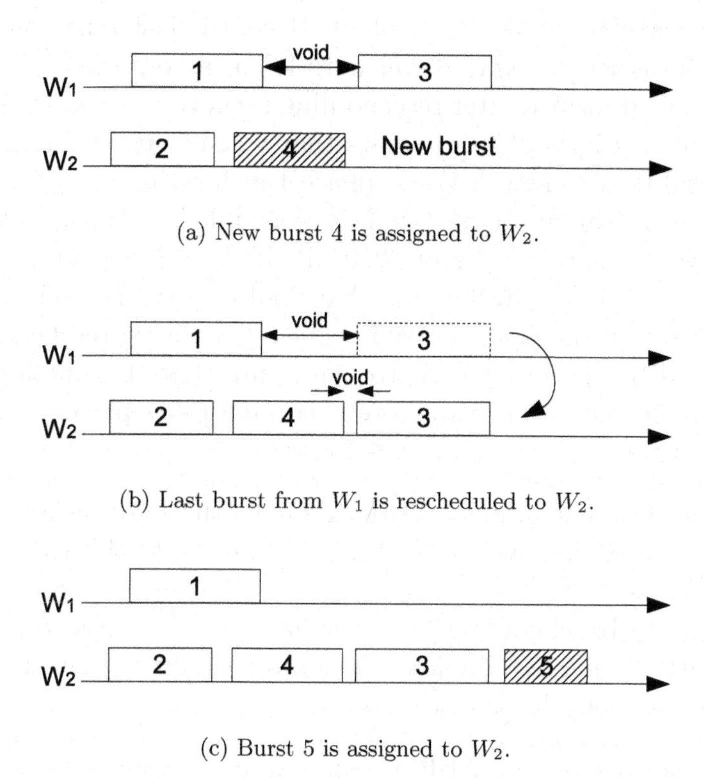

(a) New burst 4 is assigned to W_2.

(b) Last burst from W_1 is rescheduled to W_2.

(c) Burst 5 is assigned to W_2.

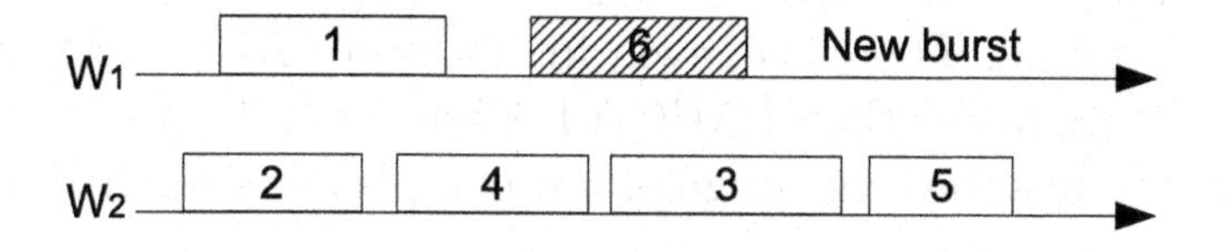

(d) Burst 6 will be able to be scheduled to W_1.

Fig. 2.10. Illustration of working of ABR algorithm.

any wavelength by LAUC, LAUC-VF, or ODBR. For the same burst arriving pattern, Figure 2.10(a) to (d) show that with ABR, the new burst 6 which would otherwise have been dropped, can be scheduled successfully. This demonstrates that prevention of burst dropping is achieved by using ABR. As shown in Figure 2.10(a), when burst 4 is scheduled to wavelength W_2, consideration for rescheduling of one last burst from other wavelength to current wavelength W_2 takes place. Here, the last burst on wavelength W_1 is scheduled to wavelength W_2 as it conforms to the rule that the void formed after rescheduling is shorter, as shown in Figure 2.10(b). Figure 2.10(c) shows that when burst 5 arrives, it is scheduled to wavelength W_2 in phase 1 as it is the latest available wavelength. Finally, burst 6 will be scheduled to W_1 at time t of its arrival as shown in Figure 2.10(d). If there are more than one burst on different wavelengths that could be rescheduled to W_p, the burst with the smallest void formed after being rescheduled to W_p would be chosen. This is to make sure that the smallest void would be formed every time a rescheduling takes place.

The complexity of ABR is as follows.

- Phase 1 - Scheduling : ABR examines the information of one burst on each wavelength. Phase 1 runs in $O(W)$ time in the worst case.
- Phase 2 - Rescheduling : Phase 2 has the worst case complexity of $O(W)$ time as well since it examines only the last burst on each wavelength for rescheduling.

The complexity of ABR is approximately two times that of LAUC since it examines only the last burst on each wavelength for rescheduling. Therefore, even for values of $K \geq 3$, ABR is expected to run faster than LAUC-VF whose complexity is $O(KW)$. It therefore has the advantage of low complexity similar to LAUC.

2.4.2 Signalling overhead

Additional signalling is needed when rescheduling is successful by using ODBR or ABR. This is to notify the next node about the change of wavelength by sending a NOTIFY packet. However, rescheduling does not incur significant signalling overhead for both ODBR and ABR algorithms. The NOTIFY packet is much smaller than the header packet as it needs to carry only the wavelength change information. Also, no complex algorithm is executed upon receiving the NOTIFY packet. Further, a successful reschedule requires a NOTIFY packet to be sent on only one link. Alternatively, without sending extra signalling packet, the information can be piggybacked to the header packet. It therefore does not incur significant processing time and does not consume significant control channel bandwidth when compared to the computational complexity gain achieved over existing algorithms such as LAUC-VF.

2.4.3 Performance study

The performance of the burst rescheduling techniques is studied via simulation in [43]. A random topology with 32 nodes and 60 bidirectional links is considered. Each link has 8 wavelengths for carrying data traffic. The transmission capacity of each wavelength is approximately 10 Gbps. No FDL buffer is assumed in the network. Network traffic consists of two classes, namely class 1 and class 2. Class 2 traffic is given a higher priority over class 1 traffic by assigning an extra offset time. The performance of ODBR and ABR is evaluated under different traffic loading conditions. The burst arrival rate is measured as the number of bursts arrived per node per microsecond. The range for traffic load is chosen to be from 0.3 to 0.6 so that the burst dropping probability is below 15%.

Figures 2.11, 2.12, and 2.13, indicate that ODBR and ABR have better performance in terms of the overall burst dropping probability and that for class 1, and class 2 traffic, respectively, than LAUC. The dropping probability increases with increasing traffic load as most of the wavelengths are heavily used at high traffic load, therefore, it is less probable for a burst to find an available wavelength. However, as shown in Figures 2.11 and 2.12, the performances of ODBR and ABR are always in between LAUC and LAUC-VF. Particularly, ODBR and ABR perform closer to LAUC-VF at low arrival rates than at high arrival rates. This is because more voids are created at high arrival rates and the rescheduling algorithms consider only the last burst for rescheduling and do not utilize the voids in between the burst as in LAUC-VF.

Figure 2.13 shows that all algorithms have similar dropping performance for class 2 high priority traffic. This is because class

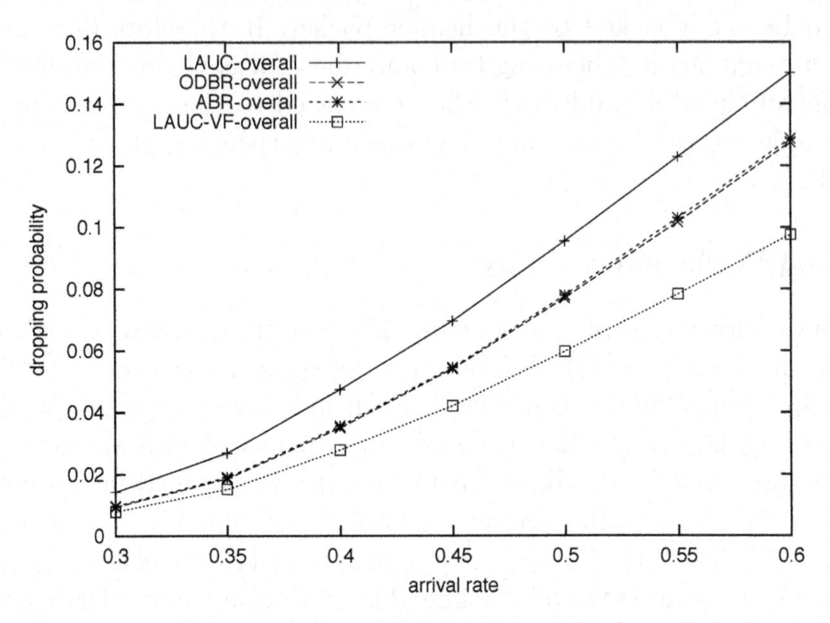

Fig. 2.11. Performance of overall traffic under different traffic loading.

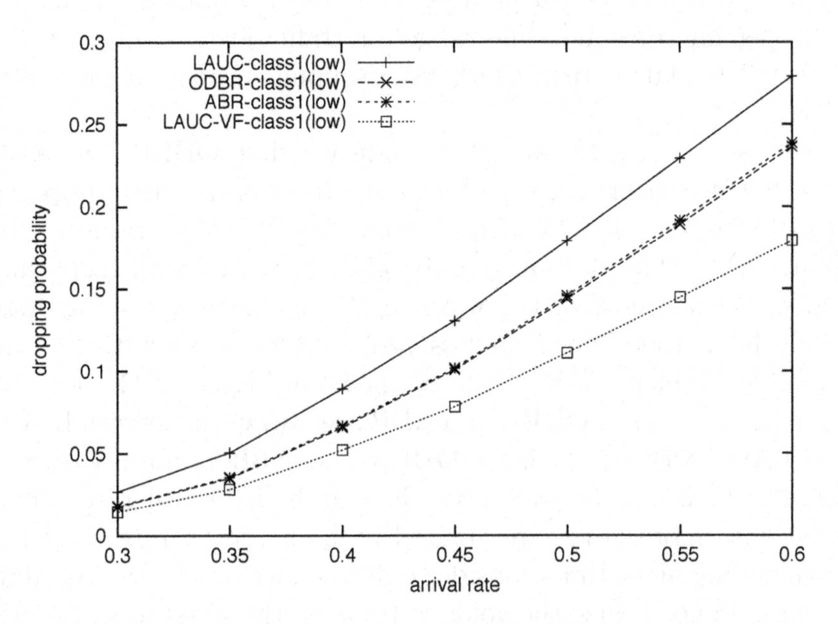

Fig. 2.12. Performance of class 1 traffic under different traffic loading.

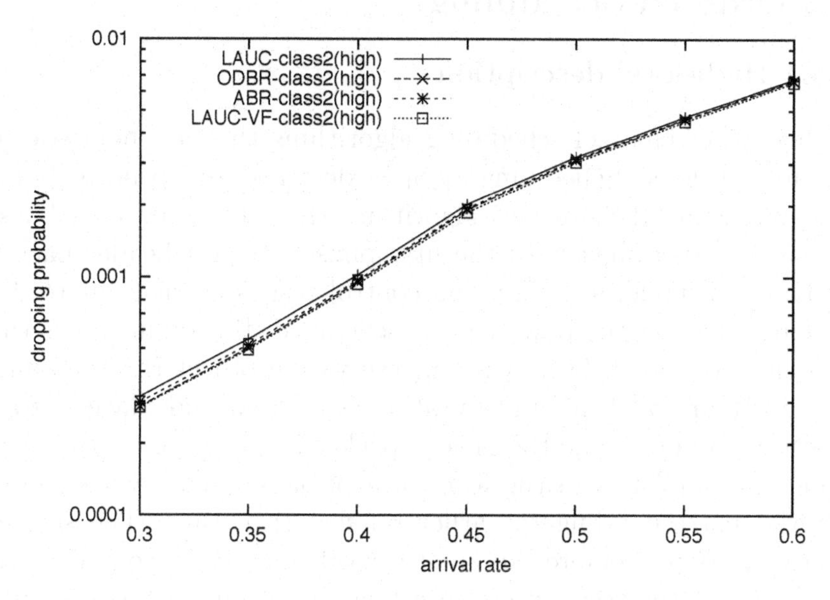

Fig. 2.13. Performance of class 2 traffic under different traffic loading.

2 traffic have large initial offset time as compared to class 1 traffic, which makes a class 2 burst highly likely to reserve wavelength at the far end on the time line. Therefore, a high priority burst is highly likely to be the last burst and hence not much improvement is achieved by LAUC-VF and the rescheduling algorithms over LAUC.

The simulation study in [43] has also observed the number of header packets that correspond to successful bursts and the number of NOTIFY packets that correspond to successful rescheduling on each of the links. The results show that about 2% and 20% of signalling overhead is incurred by ODBR and ABR, respectively.

2.5 Ordered Scheduling

2.5.1 High-level description

Unlike other channel scheduling algorithms that are heuristic in nature, Ordered Scheduling[2] [45] is designed to optimise burst scheduling in OBS. In this algorithm, the scheduling of a burst consists of two phases. In the first phase, when a header packet arrives at a node, an admission control test is carried out to determine whether the burst can be scheduled. If the burst fails the test, it is dropped. Otherwise, a reservation object that contains the burst arrival time and duration is created and placed in an electronic buffer while the header packet is passed on to the next node. The buffer is in the form of a priority queue[3] with higher priority corresponding to earlier burst arrival time. The second phase starts just before the burst arrival time. Because of the priority queue, the reservation object of the incoming burst should be at the head of the queue. It is dequeued and a free wavelength is assigned to the burst. A special NOTIFY packet is immediately generated and sent to the next downstream node to inform it of the wavelength that the burst will travel on.

A simple example shown in Figure 2.14 helps to illustrate the main concept of Ordered Scheduling. The left section of the figure shows the order of the incoming bursts and the order of the header packets in the control channel. The middle section shows that the reservation objects are placed in the priority queue in the order of burst arrivals. Finally, the scheduled bursts are shown in the right section of the figure. The node simply dequeues a reservation object from the priority queue and assigns a free wavelength to it. Since the priority queue sorts the reservations according their burst arrival times, all unscheduled reservations will be to the left of the newly dequeued reservation. Therefore, any void it produces on the right has no effect on the schedulability of non-scheduled

[2] Reprinted from (M. H. Phung, K. C. Chua, G. Mohan, M. Motani, T. C. Wong, and P. Y. Kong, "On Ordered Scheduling for Optical Burst Switching," *Computer Networks*, vol. 48, no. 6, pp. 891–909), ©(2005), with permission from Elsevier.

[3] A priority queue is a data structure that always has the highest priority element at the head of the queue.

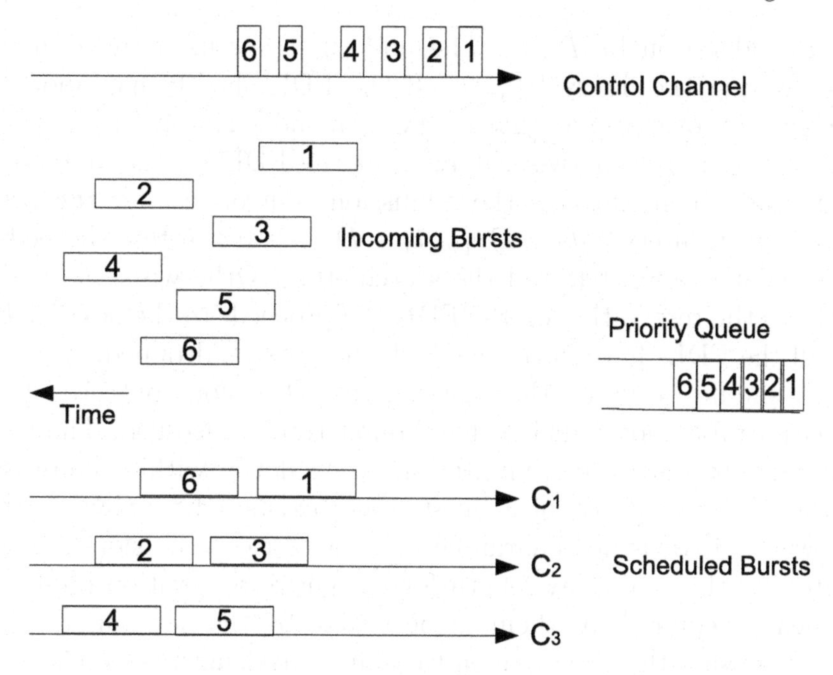

Fig. 2.14. The main concept of Ordered Scheduling

reservations. Thus, any free wavelength can be assigned to the newly dequeued reservation. In the example, a round robin assignment is used because it is the easiest way to implement.

The admission control test for an output link without an FDL buffer is given below.

A burst requesting reservation for the time interval $[t_0, t_1]$ *can be scheduled on an output link with* M *wavelengths if* $\forall t \in (t_0, t_1)$, *the number of existing reservations containing* t *is no more than* $M - 1$.

(Note: A reservation for interval $[t_0, t_1]$ is said to contain t if $t_0 < t < t_1$)

When an output link is equipped with an FDL buffer, which can be thought of as a collection of fibres (or FDLs) with different lengths, a node has the option of delaying a burst by routing it through one of the FDLs. In this case, the above admission control test is extended as follows. If a burst fails to reserve an output wavelength at its original arrival time t_0, the node searches through the FDLs in order of increasing length. Let the length of the FDL in consideration be D_{FDL}. The node first checks if the FDL is free during the interval $[t_0, t_1]$. If the FDL already has another reservation overlapping that interval, the node simply proceeds to the next FDL. Otherwise, it reserves the FDL for that interval. The node then executes the admission control test for the new reservation interval $[t_0 + D_{FDL}, t_1 + D_{FDL}]$. If the test succeeds, the burst is admitted and the search stops. Otherwise, the node undoes the reservation on the FDL and proceeds to the next FDL. If all the FDLs have been searched, the burst is dropped.

It should be noted that passing the admission control test is necessary but not sufficient for a burst to be scheduled. The test guarantees that at any infinitesimal time slot δt within the reservation interval $[t_0, t_1]$ of a burst, there exists a free wavelength. However, it does not guarantee that those free time slots are located on the same wavelength for the entire reservation interval, which is required for them to be usable by the new burst. The key to ensure that they are on the same wavelength is to schedule bursts in the order of their arrival times as is done in the second phase using the priority queue.

It can be seen that Ordered Scheduling is similar to Dual-header OBS described in section 2.3. The primary difference between the two schemes is the admisison control test. The admission control test is important because it prevents resource wastage due to over-admitting bursts, also known as phantom reservations in Dual-header OBS. In both schemes, header packets are passed on to the next node before scheduling takes place. Thus, without the admission control test, an incorrectly admitted burst will have its header packet forwarded to downstream nodes to make further reservations. However, the node that makes the incorrect admission will not be able to schedule the burst. Without having the optical switch configured for it, the burst will be lost upon arrival and resources reserved at downstream nodes will be wasted.

Ordered Scheduling is optimal in the sense that given some previously admitted burst reservations, if an incoming burst reservation cannot be admitted by Ordered Scheduling, it cannot be scheduled by any other scheduling algorithm. This is because by definition, if a new burst reservation fails the admission control test, there exists a time slot within its reservation interval in which all the data wavelengths are occupied. Therefore, the only way to schedule the new burst is to preempt some existing reservations.

2.5.2 Admission control test realisation

The admission control test in section 2.5.1 is presented in continuous form, which may not be practical or feasible to realise. A simple solution would be to divide the time axis into slots. A burst that reserves any portion of a time slot, however small, will be considered as occupying the whole time slot. The admission control routine simply needs to keep track of the number of admitted bursts $N_{occupied}$ that occupy each time slot and compare it to the total number of data wavelengths M. A new burst will be admitted only if $N_{occupied} < M$ for all the slots it will occupy. This version is referred to in [45] as Basic Ordered Scheduling. It is suitable if the optical switches in the network also operate in a slotted fashion as mentioned in [32]. In that case, the time slots chosen by Ordered Scheduling should simply be set to be the same as the time slots of the underlying optical switches.

The basic slotted approach, however, may degrade the system performance if the underlying optical switches can operate in a truly asynchronous fashion. Due to its discrete nature, it does not consider the case where two bursts can occupy the same wavelength in a slot and thus may lead to unnecessary burst loss. This may be alleviated by having the slot size much smaller than the average burst size. However, that will increase the processing time and/or hardware complexity.

An enhanced version of the above slotted approach is called Enhanced Ordered Scheduling in [45]. Instead of a single number to indicate the number of bursts occupying a time slot, the admission control routine keeps the following three data entities for each time slot:

1. N_{total} is the total number of bursts that occupy the slot, whether wholly or partly;

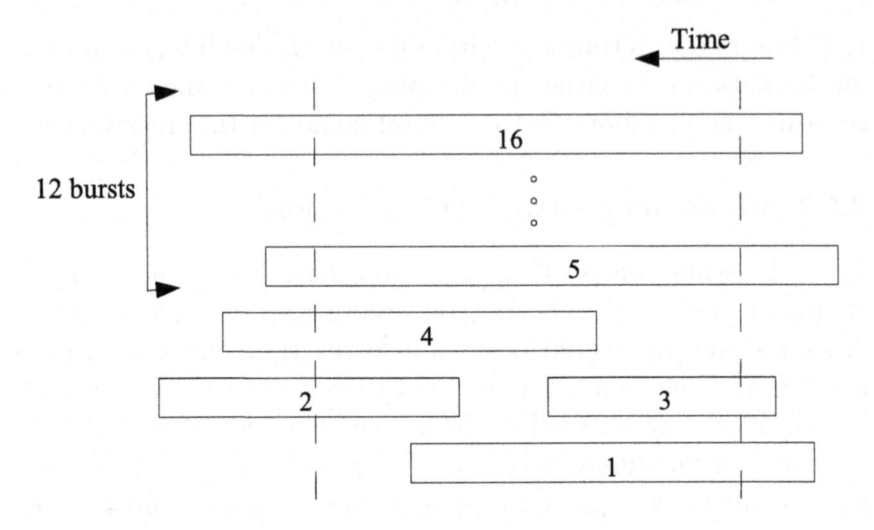

Fig. 2.15. Example of the bookkeeping for a typical time slot.

2. *heads* is the list of the start times of the bursts that have the start of their reservation periods fall within the slot, sorted in increasing order; and
3. *ends* is the list of the end times of the bursts that have the end of their reservation periods fall within the slot, sorted in increasing order.

The bookkeeping of the two versions is illustrated in Figure 2.15. In the figure, the slot sizes are exaggerated and the bursts not shown occupy the entire slot. For the basic version, $N_{occupied} = 16$. For the enhanced version, $N_{total} = 16$ and there are two entries in each of *heads* and *ends*.

When a header packet arrives at a node to request a reservation, the admission control routine pretends that the burst has passed the test and updates the database of all the time slots involved, i.e., the time slots of the burst corresponding to the arriving header packet. N_{total} is incremented by one for each of the time slots. In addition, for the time slots containing the start or the end of the burst, an entry is added to *heads* or *ends*, respectively. The admission control routine then checks all the involved time slots. For a particular slot, if N_{total} is larger than the number of wavelengths M, entries in the *heads* list will be *matched* to those in the *ends* list to reduce the number of occupied wavelengths. A pair of bursts are considered matched for a given slot if the start time of the one in the *heads* list is no greater than the end time of the other in the *ends* list. The actual number of occupied wavelengths is N_{total} minus the number of matched pairs. If this number is smaller than M for all the involved slots, the new burst is schedulable and admitted. Otherwise, its header packet is dropped and its information previously inserted in the time slots' database is removed.

The matching operation is facilitated by the fact that *heads* and *ends* are kept in increasing order. The algorithm simultaneously goes from the beginning to the end on both lists, checking their entries against each other. Let i and j be the current indices on *heads* and *ends*. If $heads[i] \leq ends[j]$ then a match is recorded and i and j are incremented by one. Otherwise, only j is incremented to point to the next larger entry in *ends*. The process is repeated until either i or j passes the end of the list.

The formal description of the algorithm is presented in Table 2.3. Denote $[t_0, t_1]$ as the requested reservation interval; *slot* as the object representing a particular time slot and s_0 and s_1 as the slots that contain t_0 and t_1, respectively. Also, let $slot.insert\_head(t)$ and $slot.insert\_end(t)$ be the functions that insert t into the sorted lists *heads* and *ends* of *slot*, respectively. The main test procedure uses three sub-functions $insert(t_0, t_1)$, $match()$ and $remove(t_0, t_1)$. The first two sub-functions are presented below the main test procedure while the last one is omitted because it is similar to $insert(t_0, t_1)$.

In terms of loss performance, Enhanced Ordered Scheduling is optimal since it fully implements the test in continuous form. Its outperformance compared to the basic version is illustrated in Figure 2.15 where the basic version reports that 16 wavelengths are occupied for the time slot while the enhanced version reports only 15 occupied wavelengths. The disadvantage of the enhanced version is that it is more complex. This will be explored in the next section.

Table 2.3. Admission control test

```
MAIN()
accept ← true
for slot = s0 to s1
    slot.insert(t0, t1)
    if slot.N_total − slot.match() > M
        accept ← false
if accept = false
    for slot = s0 to s1
        slot.remove(t0, t1)

INSERT(t0, t1)
N_total ← N_total + 1
if slot contains t0 then insert_head(t0)
if slot contains t1 then insert_head(t1)

MATCH()
i ← 0
j ← 0
matched ← 0
while Neither i nor j have passed the end of heads and ends, respectively
    if heads[i] ≤ ends[j]
        matched ← matched + 1
        i ← i + 1
    j ← j + 1
return matched
```

2.5.3 Complexity analysis

Admission control test

The slotted structure of the two admission control implementations is particularly suitable for parallel processing since each slot of a burst is processed independently. Let S be the maximum number of slots in the scheduling window. A simple parallel solution is to have S processing elements in the admission control unit with each processing element responsible for one slot. When a header packet arrives, the processing elements corresponding to the slots covered by the burst will compute the admission control test simultaneously.

The time complexity analysis for the basic and enhanced versions of the admission control test is as follows. For Basic Ordered Scheduling, a processing element needs to perform at most one comparison and one update of $N_{occupied}$ per burst. Therefore, the required processing time is constant and takes less than 1 ns assuming a processing speed in the order of 10^9 operations per second. For Enhanced Ordered Scheduling, the processing element also needs to perform one comparison and one update. In addition, it needs to do the matching operation when necessary. Assuming that the slot size is smaller than the minimum burst size, the number of elements in *heads* and *ends* is M in the worst case. So the worst case complexity of the matching operation is $O(M)$. Also, the update of *heads* and *ends* at the two slots at the two ends of a burst takes $O(logM)$. Therefore, the overall worst case time complexity is $O(1)+O(M)+O(logM) = O(M)$. In a normal case, however, the size of *heads* and *ends* is about M/K where K is the average number of slots per burst. Hence, the average complexity is $O(M/K)$ per matching operation. The overall average complexity is $O(1) + O(M/K) + O(logM) = O(M/K + logM)$. As an example, let $M = 256$ and $K = 16$; *heads* and *ends* will have about 16 elements on average. A worst case estimate of the processing time is 50 ns, which includes the execution of $match()$ and $remove(t_0, t_1)$. The average processing time is much smaller as $match()$ and $remove(t_0, t_1)$ are only executed in heavy loading conditions.

The required number of processing elements is inversely proportional to the slot size, or proportional to the average number of slots per burst K. Therefore, although Basic Ordered Scheduling has the advantage of fast processing compared to the enhanced version, its drawback is that it requires a much larger number of processing elements to ensure good burst dropping performance. For Enhanced Ordered Scheduling, there is a tradeoff between processing speed and hardware complexity. A small value of K will reduce the required number of processing elements but will lead to longer execution time and vice versa.

For comparison, it is possible to perform a parallel search across the wavelengths to find all the unused wavelengths for LAUC-VF. Then the search results are compared to each other to find the latest available one. These operations can be performed in $O(logM)$ time, which is better than Enhanced Ordered Scheduling and worse than Basic Ordered Scheduling. In terms of hardware complexity, LAUC-VF requires one processing element for each wavelength with each processing element being fairly complex. If the number of wavelengths per link is large, which is usually the case, the hardware requirement for LAUC-VF will be larger than that for Ordered Scheduling.

Priority queue

The queueing operations on the priority queue are common to both versions of Ordered Scheduling. Its complexity depends on the specific implementation of the underlying priority queue. Some efficient implementations of priority queues using pipelined heap are reported in the literature [46, 47]. They have $O(1)$ time complexity with regard to queue size. Implemented on conservative technologies such as 0.35-micron and 0.18-micron CMOS, they can achieve up to 200 million queueing operations per second for queues with up to 2^{17} entries. The queue size depends on the size of the scheduling window, which in turn depends on offset times and FDL buffer depth, and the burst arrival rate. Note that the above priority queue implementations can accommodate any queue size of practical interest. The queue size only affects the amount of required memory.

Overall time complexity

The computational work in admission control and priority queue operations can be pipelined. That is, as soon as the admission control routine finishes with a header packet and passes it to the priority queue, it can handle the next header packet while the first header packet is being enqueued. Therefore, the overall complexity is the maximum of the two parts. With parallel processing, the time complexity for Basic Ordered Scheduling is $O(1)$. The worst case and average time complexities for Enhanced Ordered Scheduling are $O(M)$ and $O(M/K + logM)$, respectively.

2.5.4 Performance study

In [45], a simulation model is used to evaluate the performance of the Ordered Scheduling algorithms. Both versions of Ordered

Scheduling are investigated. The slot sizes are 1 μs and 0.1 μs for the enhanced and basic versions, respectively, unless otherwise stated. The reason for the difference in the chosen slot sizes is because the performance of Basic Ordered Scheduling critically depends on the slot size while the performance of Enhanced Ordered Scheduling does not. LAUC-VF operating under the same condition is used for comparison.

The burst assembly algorithm is a simple time-based algorithm with a time limit T_{limit}. There are separate assembly queues for each ingress node and incoming IP packets choose a queue with equal probability. When the first IP packet that forms a burst arrives at an assembly queue, a timer is started from zero. Subsequent IP packets are appended to the assembly queue. A burst will be created when the timer exceeds T_{limit}. In the experiments, T_{limit} is set such that the maximum burst duration is 2.5 μs. So the average number of slots per burst are approximately 2 and 20 for Enhanced and Basic Ordered Scheduling, respectively.

The simulation study consists of three sets of experiments. The first two sets are carried out for a topology with a single core node. They aim to investigate the effects of traffic conditions and hardware configurations on the performance of the algorithms, respectively. The final experiment set is carried out for an entire network to investigate the effect of network topology on the performance trend among the algorithms.

The simulation topology for the first two sets of experiments is shown in Figure 2.16. There are four ingress nodes and four burst sinks connected to the core node. The burst sinks represent egress nodes in a real network. No burst dropping is assumed on the links between the ingress nodes and the core node. It only occurs on the output links of the core node.

Effects of traffic conditions

In this set of experiments, the configuration is as follows. The links connecting the OBS nodes are made up of a single optical fibre per link. Each optical fibre has 8 data wavelengths. The core node has an FDL buffer with 6 FDLs of lengths 5 μs, 10 μs,. . . , 30 μs.

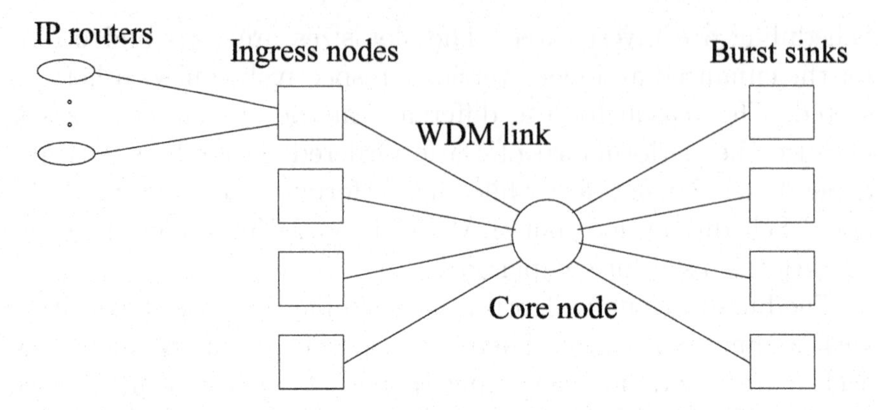

Fig. 2.16. Topology for simulation study of a single core node

Firstly, the effect of varying the offered load to the core node is examined. For this experiment, two offset-based QoS classes with equal loading are used. Class 2 is given higher priority than class 1 by assigning an extra offset of 3 μs. This offset difference is larger than the maximum burst size so that full isolation between the two classes is achieved. The arrival rate ranges from 3.3 bursts per μs to 4.15 bursts per μs, or from 0.74 to 0.92 in terms of offered load. Offered loads lower than 0.74 are not considered because they would make the loss probability of class 2 too small to measure through simulation. On the other hand, offered loads larger than 0.92 would make the loss probability of class 1 too large to be of practical interest.

The simulation results are plotted in Figure 2.17. They show that the burst dropping probabilities increase with increasing offered load, which is expected. Among the algorithms, Enhanced Ordered Scheduling has the best burst dropping performance followed by Basic Ordered Scheduling and then LAUC-VF. This order of burst dropping performance among the algorithms is as expected based on the discussion in the previous sections. The order of performance is the same in virtually all of the following experiments. Note that the differences in performance are greater at lower load. This is because at low load, there are more free wavelengths to choose from to assign to an incoming burst reservation and LAUC-VF is more likely to make suboptimal wavelength as-

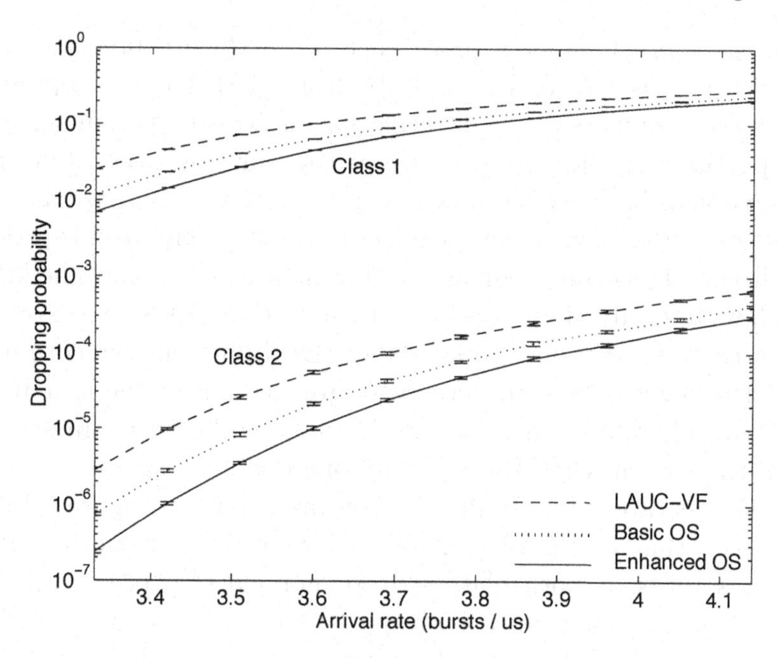

Fig. 2.17. Burst loss probability versus traffic loading

signment decisions due to incomplete knowledge of other burst reservations. Between the two classes, it is observed that the performance improvement of Ordered Scheduling over LAUC-VF is greater for class 2 than it is for class 1. The reason for this is also related to loading. Since full isolation is achieved between the two classes, the effective loading for class 2 traffic is only half of that for class 1 traffic. So as the above reasoning goes, the improvement for class 2 is larger.

The effect of traffic class composition is considered next. The same traffic parameters as above are used except that the overall offered load is fixed at 0.9 and the offered load of each class is varied. As the proportion of class 1 traffic varies from 0 to 1, Figure 2.18(a) shows that the overall traffic loss rate follows a bell-shaped curve, which is slightly tilted towards the left. The burst dropping probabilities peak when the burst rates from the two classes are comparable and are at the lowest at the two extremities where traffic from only one class is present. This effect can be explained from the queueing model point of view. As the traffic composition becomes more balanced, more class 1 bursts are preempted by those from class 2. When burst B_1 from class 1 is preempted by burst B_2 from class 2, the effective size of B_2 is its actual size plus the portion already served of B_1. If the burst size distribution is not exponential, that will increase the effective burst size and the burst loss rates. This negative effect of burst preemption is present in all the algorithms, unlike the fragmentation of the scheduling window that only affects LAUC-VF. Note that at the two extremes when there is only one traffic class, FDL buffers in the core node can still delay bursts and create fragmentation in the scheduling window. Therefore, there are performance differences among the algorithms even when there is only one traffic class.

The loss rates of individual classes are shown in Figure 2.18(b). It is seen that as the proportion of low priority traffic increases, the loss rates of both classes drop. For class 1, preemption by class 2 bursts make up a large part of its burst loss. Therefore, when there is less class 2 traffic, preemption occurs less frequently, which leads to the drop in class 1 burst loss. For class 2, the only cause for burst loss is intra-class contention since it is fully isolated from class 1. Thus, when its traffic rate decreases, contention rate rapidly decreases and so does the burst loss. This result implies that very low burst dropping probability can be achieved for high priority traffic even though the overall utilisation is high.

The final traffic parameter to be investigated is the number of QoS classes. In this experiment, the overall offered load is 0.8 and all traffic classes have equal loading. The overall burst dropping probabilities are plotted in Figure 2.19. It shows that the overall burst dropping probability increases as the number of classes increases. This is as expected because as the number of classes increases, the scheduling window is more fragmented, which results in increasing loss probability. A notable aspect is the large increase in loss probabilities moving from one to two classes. This is caused by the large increase in the degree of fragmentation in the scheduling window when moving from one class to two classes.

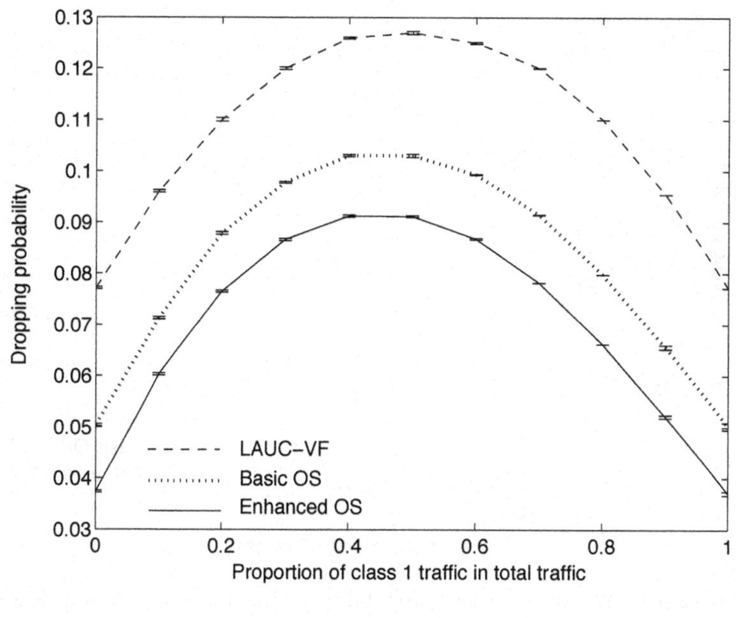

(a) Overall performance

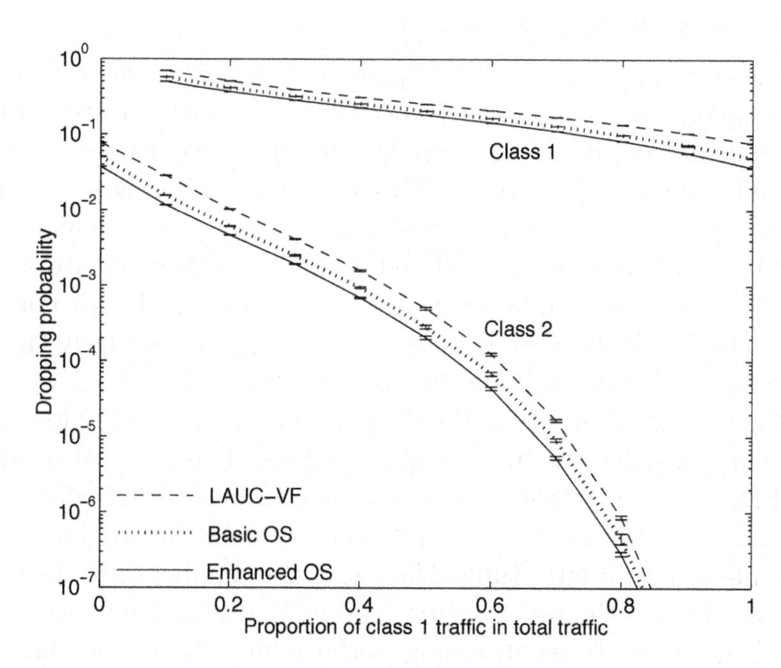

(b) Performance of individual traffic classes

Fig. 2.18. Effect of traffic composition on burst loss probability

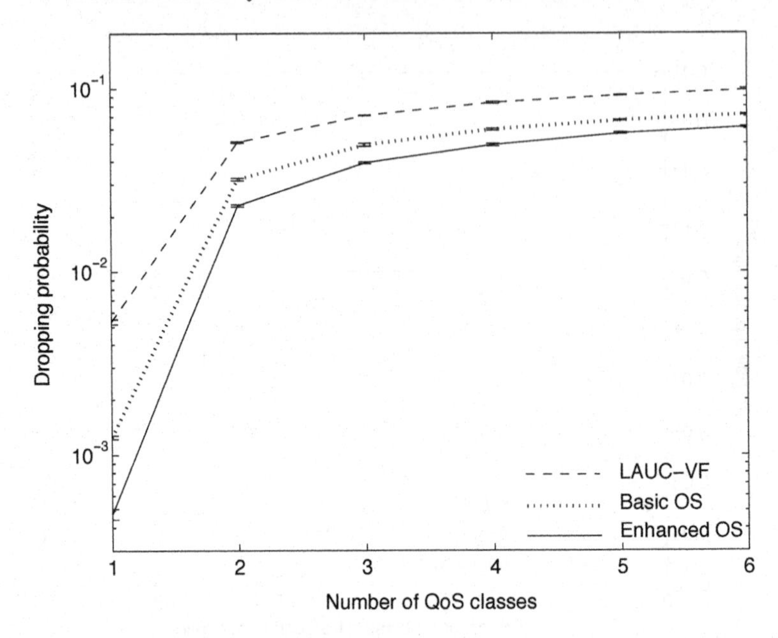

Fig. 2.19. Overall burst loss probability versus number of traffic classes

Effects of hardware configuration

In the first experiment, the impact of FDL buffer depth on the performance of the algorithms is studied. Two kinds of data traffic are considered: one with a single QoS class and the other with two offset-based QoS classes. This is because the effects on burst dropping performance are slightly different between having one and two QoS classes. The offered loads for both cases are set at 0.8. The FDL buffer in use consists of a number of FDLs according to the buffer depth. The lengths of the FDLs are regularly spaced starting from 5 μs with length spacing being 5 μs.

Figure 2.20 shows that the overall trend is improving loss performance with increasing number of FDLs. This is because when an FDL buffer is introduced, if a node cannot schedule a burst at its original arrival time, the node can try delaying the burst and scheduling it at a later time. The larger the number of FDLs there are in a buffer, the more options the node has in delaying bursts, which improves burst dropping performance. Note also that the curves for LAUC-VF tend to level off. This can be explained by the fact that the scheduling window is increasingly fragmented as more FDLs are introduced. For LAUC-VF, this negative effect opposes and neutralises the beneficial effect of having more FDLs, which explains the levelling off of its curve. Ordered Scheduling, on the other hand, is not affected due to its deferment of scheduling decisions. The above effect is more pronounced in Figure 2.20(a) than it is in Figure 2.20(b) because having two offset-based QoS classes already introduces significant fragmentation in the scheduling window so the additional fragmentation caused by more FDLs has less effect.

The impact of slot size on the performance of Basic Ordered Scheduling is studied next. For this and the remaining experiments, input traffic with two QoS classes is used. In this experiment, the loss performance of Basic Ordered Scheduling with different slot sizes is measured at an overall offered load of 0.6 and compared to Enhanced Ordered Scheduling and LAUC-VF. The results are plotted in Figure 2.21. Since the performance of the latter two algorithms is not affected by slot size, their loss curves show up as horizontal lines. On the other hand, as the slot size gets larger, the burst dropping performance of Basic Ordered Scheduling rapidly worsens due to its discrete implementation of the admission control test. At a slot size of 1 μs, which is what is used by Enhanced Ordered Scheduling, the dropping probability of Basic Ordered Scheduling is nearly three orders of magnitude larger than Enhanced Ordered Scheduling. These results confirm the necessity to use much smaller slot sizes for Basic Ordered Scheduling compared to Enhanced Ordered Scheduling.

The final experiment in this section investigates the effects of the number of wavelengths per link on the performance of the algorithms. The performance results of a non-void filling scheduling scheme as described in section 2.2.1 is included. The purpose is to see how its performance compares to those of other void filling algorithms at different numbers of wavelengths. The burst dropping probabilities are measured at an overall offered load of 0.8 and different numbers of wavelengths per link and plotted in Figure 2.22. It shows that the overall trend is decreasing loss probabilities with increasing number of wavelengths per link. This is the direct result of an OBS switch behaving like an $M|M|k|k$ loss system. Observe

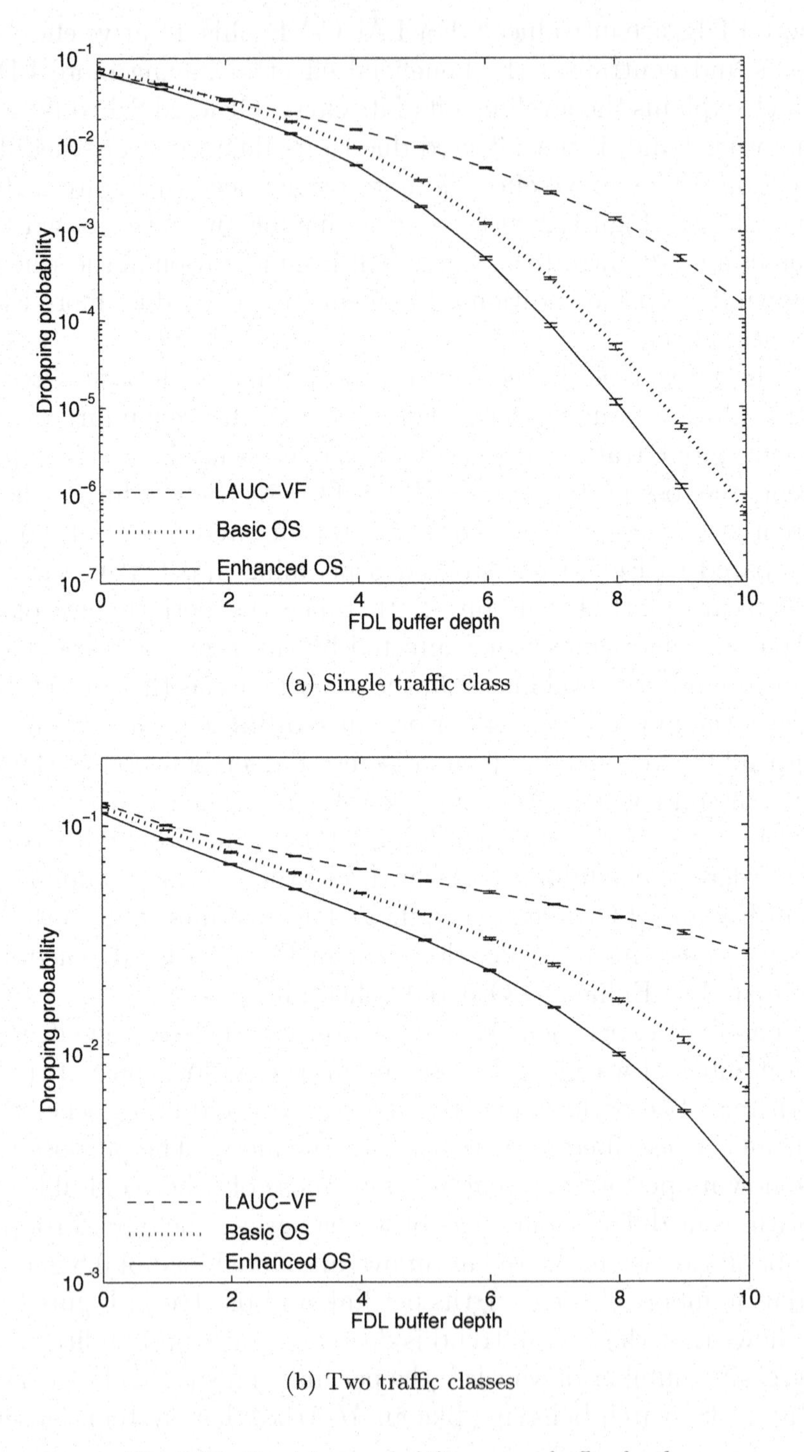

(a) Single traffic class

(b) Two traffic classes

Fig. 2.20. Burst loss probability versus buffer depth

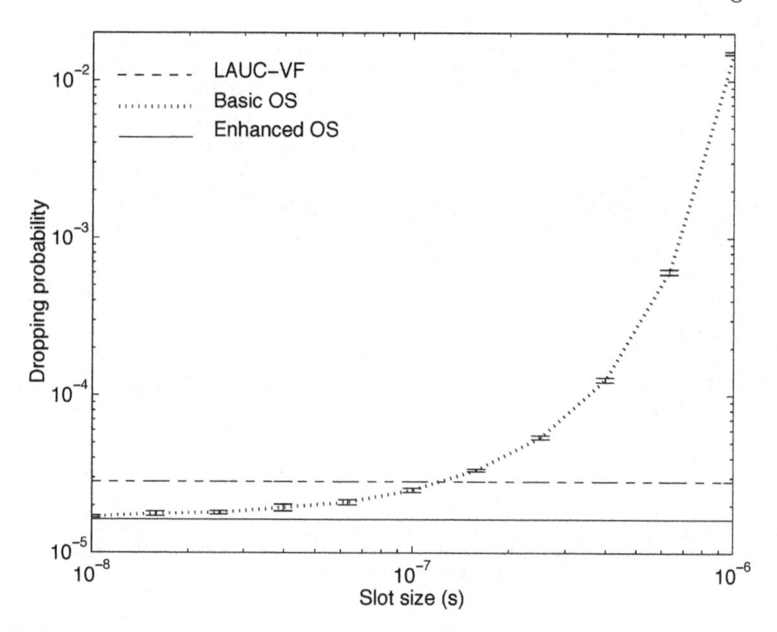

Fig. 2.21. Performance of Basic Ordered Scheduling with different slot size

also that the performance of the Horizon scheme is poor when the number of wavelengths is low but gets very close to that of LAUC-VF when the number of wavelengths is high. Among the void filling algorithms, the relative performance between Enhanced Ordered Scheduling and LAUC-VF remains the same. However, the relative performance of Basic Ordered Scheduling compared to the enhanced version gradually decreases as the number of wavelengths per link increases. This is also due to the discrete nature of Basic Ordered Scheduling. As a slot handles more and more bursts, the chance that Basic Ordered Scheduling over-reports the number of occupied wavelengths as illustrated in Figure 2.15 increases. From this experiment and the previous one, it is observed that the performance of Ordered Scheduling depends on the ratio between the number of slots per burst and the number of wavelengths per link.

Simulation study for an entire network

The three scheduling algorithms are next simulated in a realistic network setting to see if the network topology affects the performance trend among the algorithms. For this experiment, the

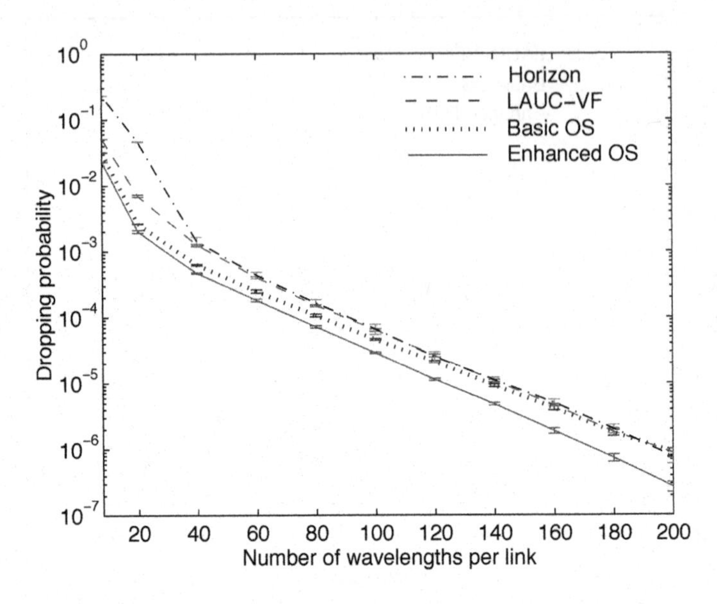

Fig. 2.22. Burst loss probability versus number of wavelengths per link

topology in Figure 2.23, which is a simplified topology of the US backbone network, is used. The topology consists of 24 nodes and 43 links. The average node degree is 3.6. Shortest path routing is used to determine the transmission paths among nodes and the average hop length of the paths is 3. For simplicity, the propagation delays between adjacent nodes are assumed to have a fixed value of 10 ms. The links are bi-directional, each implemented by two uni-directional links in opposite directions.

The input traffic and hardware configuration for each node remain the same, i.e., two offset-based QoS classes, eight wavelengths per link and six FDLs per output link at each node. Each node has 23 separate assembly queues, one for every other node. An incoming IP packet to a node enters one of the queues with equal probability. The IP packet arrival rates are the same for every node. The header packet processing time per node is assumed to be $\Delta = 1$ μs. For a path with H hops, the initial offset times assigned for bursts of classes 1 and 2 are $\Delta \cdot H$ μs and $\Delta \cdot H + 3$ μs, respectively.

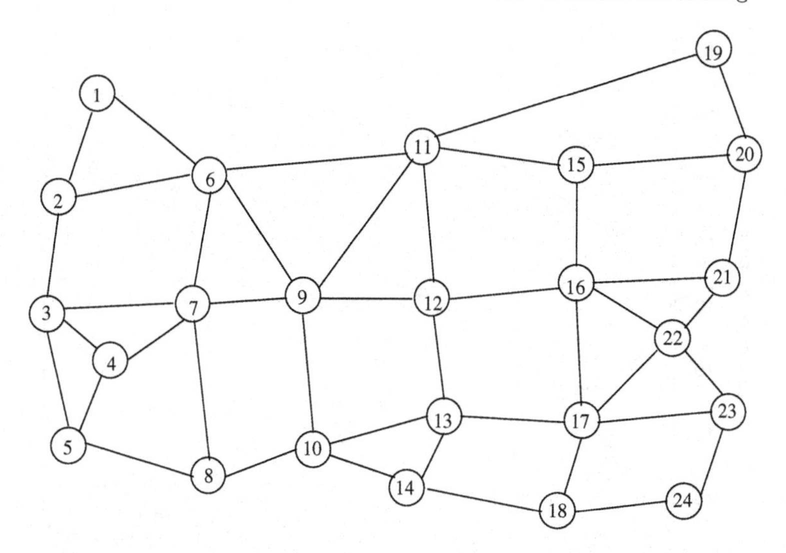

Fig. 2.23. 24-node NSF network topology

The burst dropping probabilities for the algorithms are plotted against the offered load to the network. The offered load is measured in terms of the number of departing bursts per node per μs. The simulation results are shown in Figure 2.24. It is observed that the performance trend is similar to that in Figure 2.17. The order among the algorithms remains the same, i.e., Enhanced Ordered Scheduling has the best performance, followed by Basic Ordered Scheduling and LAUC-VF. Note that the arrival rates used in this experiment are much smaller than those used in Figure 2.17 but the ranges of the burst dropping probabilities are approximately the same. This is because in a network environment, many paths may converge at some nodes, causing bottlenecks. The offered loads to those bottlenecked nodes are much larger than the average offered load to the network and most of the burst loss in the network is concentrated there.

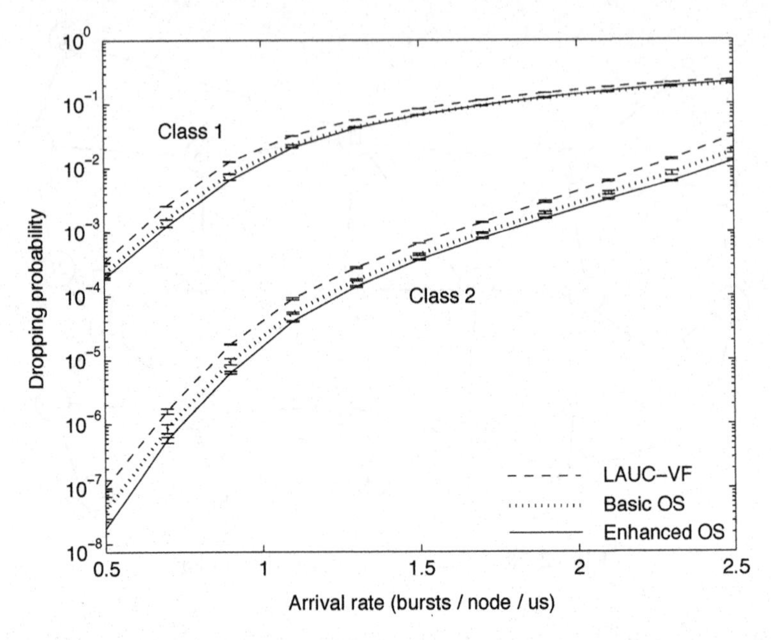

Fig. 2.24. Average burst loss probability for NSFNET versus average network load

References

1. D. B. Sarrazin, H. F. Jordan, and V. P. Heuring, "Fiber Optic Delay Line Memory," *Applied Optics*, vol. 29, no. 5, pp. 627–637, 1990.
2. I. Chlamtac *et al.*, "CORD: Contention Resolution by Delay Lines," *IEEE Journal on Selected Areas in Communications*, vol. 14, no. 5, pp. 1014–1029, 1996.
3. D. K. Hunter, M. C. Chia, and I. Andonovic, "Buffering in Optical Packet Switches," *IEEE/OSA Journal of Lightwave Technology*, vol. 16, no. 12, pp. 2081–2094, 1998.
4. W. D. Zhong and R. S. Tucker, "A New Wavelength-Routed Photonic Packet Buffer Combining Traveling Delay Lines with Delay Line Loops," *IEEE/OSA Journal of Lightwave Technology*, vol. 19, no. 8, pp. 1085–1092, 2001.
5. T. Zhang, K. Lu, and J. P. Jue, "Shared Fiber Delay Line Buffers in Asynchronous Optical Packet Switches," *IEEE Journal on Selected Areas in Communications*, vol. 24, no. 4, pp. 118–127, 2006.
6. K. K. Merchant *et al.*, "Analysis of an Optical Burst Switching Router With Tunable Multiwavelength Recirculating Buffers," *IEEE/OSA Journal of Lightwave Technology*, vol. 23, no. 10, pp. 3302–3312, 2005.
7. D. K. Hunter, W. D. Cornwell, T. H. Gilfedder, A. Franzen, and I. Andonovic, "SLOB: A Switch with Large Optical Buffers for Packet Switching," *IEEE/OSA Journal of Lightwave Technology*, vol. 16, no. 10, pp. 1725–1736, 1998.
8. I. Chlamtac, A. Fumagalli, and C. J. Shu, "Multibuffer Delay Line Architectures for Efficient Contention Resolution in Optical Switching Nodes," *IEEE Transactions on Communications*, vol. 48, no. 12, pp. 2089–2098, 2000.
9. N. Ogashiwa, H. Harai, N. Wada, F. Kubota, and Y. Shinoda, "Multi-Stage Fiber Delay Line Buffer in Photonic Packet Switch for Asynchronously Arriving Variable-Length Packets," *IEICE Transactions on Communications*, vol. E88-B, no. 1, pp. 258–265, 2005.
10. R. S. Tucker, P. C. Ku, and C. J. Chang-Hasnain, "Slow-Light Optical Buffers: Capabilities and Fundamental Limitations," *IEEE/OSA Journal of Lightwave Technology*, vol. 23, no. 12, pp. 4046–4066, 2005.
11. A. S. Acampora and S. I. A. Shah, "Multihop Lightwave Networks: A Comparison of Store-and-Forward and Hot-Potato Routing," *IEEE Transactions on Communications*, vol. 40, no. 6, pp. 1082–1090, 1992.
12. A. G. Greenberg and B. Hajek, "Deflection Routing in Hypercube Networks," *IEEE Transactions on Communications*, vol. 40, no. 6, pp. 1070–1081, 1992.

13. F. Forghieri, A. Boroni, and P. R. Prucnal, "Analysis and Comparison of Hot-Potato and Single-Buffer Deflection Routing in Very High Bit Rate Optical Mesh Networks," *IEEE Transactions on Communications*, vol. 43, no. 1, pp. 88–98, 1995.
14. T. Chich, J. Cohen, and P. Fraigniaud, "Unslotted Deflection Routing: A Practical and Efficient Protocol for Multihop Optical Networks," *IEEE/ACM Transactions on Networking*, vol. 9, no. 1, pp. 47–59, 2001.
15. S. Yao, B. Mukherjee, S. J. B. Yoo, and S. Dixit, "A Unified Study of Contention-Resolution Schemes in Optical Packet-Switched Networks," *IEEE/OSA Journal of Lightwave Technology*, vol. 21, no. 3, pp. 672–683, 2003.
16. C.-F. Hsu, T.-L. Liu, and N.-F. Huang, "Performance Analysis of Deflection Routing in Optical Burst-Switched Networks," in *Proc. IEEE Infocom*, 2002, pp. 66–73.
17. S. Lee, K. Sriram, H. Kim, and J. Song, "Contention-Based Limited Deflection Routing Protocol in Optical Burst-Switched Networks," *IEEE Journal on Selected Areas in Communications*, vol. 23, no. 8, pp. 1596–1611, 2005.
18. N. Ogino, and H. Tanaka, "Deflection Routing for Optical Bursts Considering Possibility of Contention at Downstream Nodes," *IEICE Transactions on Communications*, vol. E88-B, no. 9, pp. 3660–3667, 2005.
19. C. Cameron, A. Zalesky, and M. Zukerman, "Prioritized Deflection Routing in Optical Burst Switching," *IEICE Transactions on Communications*, vol. E88-B, no. 5, pp. 1861–1867, 2005.
20. A. Detti, V. Eramo, and M. Listanti, "Performance Evaluation of a New Technique for IP Support in a WDM Optical Network: Optical Composite Burst Switching (OCBS)," *IEEE/OSA Journal of Lightwave Technology*, vol. 20, no. 2, pp. 154–165, 2002.
21. V. M. Vokkarane and J. P. Jue, "Burst Segmentation: An Approach for Reducing Packet Loss in Optical Burst Switched Networks," *SPIE/Kluwer Optical Networks*, vol. 4, no. 6, pp. 81–89, 2003.
22. V. M. Vokkarane, , G. P. Thodime, V. U. Challagulla, and J. P. Jue, "Channel Scheduling Algorithms using Burst Segmentation and FDLs for Optical Burst-Switched Networks," in *Proc. IEEE International Conference on Communications*, 2003, pp. 1443–1447.
23. W. Tan, S. Wang, and L. Li, "Burst Segmentation for Void-Filling Scheduling and Its Performance Evaluation in Optical Burst Switching," *Optics Express*, vol. 12, no. 26, pp. 6615–6623, 2004.
24. Z. Rosberg, H. L. Vu, M. Zukerman, and J. White, "Performance Analyses of Optical Burst Switching Networks," *IEEE Journal on Selected Areas in Communications*, vol. 21, no. 7, pp. 1187–1197, 2003.
25. V. Eramo and M. Listanti, "Packet Loss in a Bufferless Optical WDM Switch Employing Shared Tunable Wavelength Converters," *IEEE/OSA Journal of Lightwave Technology*, vol. 18, no. 12, pp. 1818–1833, 2000.
26. V. Eramo, M. Listanti, and P. Pacifici, "A Comparison Study on the Number of Wavelength Converters Needed in Synchronous and Asynchronous All-Optical Switching Architectures," *IEEE/OSA Journal of Lightwave Technology*, vol. 21, no. 2, pp. 340–355, 2003.
27. M. Yao, Z. Liu, and A. Wen, "Accurate and Approximate Evaluations of Asynchronous Tunable Wavelength Converter Sharing Schemes in Optical Burst Switched Networks," *IEEE/OSA Journal of Lightwave Technology*, vol. 23, no. 10, pp. 2807–2815, 2005.

28. G. Shen, S. K. Bose, T. H. Cheng, C. Lu, and T. Y. Chai, "Performance Study on a WDM Packet Switch with Limited-Range Wavelength Converters," *IEEE Communications Letters*, vol. 5, no. 10, pp. 432–434, 2001.
29. Z. Zhang and Y. Yang, "Performance Modeling of Bufferless WDM Packet Switching Networks with Wavelength Conversion," in *Proc. IEEE Globecom*, 2003, pp. 2498–2502.
30. V. Eramo, M. Listanti, and M. Spaziani, "Resources Sharing in Optical Packet Switches with Limited-Range Wavelength Converters," *IEEE/OSA Journal of Lightwave Technology*, vol. 23, no. 2, pp. 671–687, 2005.
31. J. S. Turner, "Terabit Burst Switching," *Journal of High Speed Network*, vol. 8, no. 1, pp. 3–16, 1999.
32. Y. Xiong, M. Vandenhoute, and H. C. Cankaya, "Control Architecture in Optical Burst-Switched WDM Networks," *IEEE Journal on Selected Areas in Communications*, vol. 18, no. 10, pp. 1838–1851, 2000.
33. L. Tančevski, S. Yegnanarayanan, G. Castanon, L. Tamil, F. Masetti, and T. McDermott, "Optical Routing of Asynchronous, Variable Length Packets," *IEEE Journal on Selected Areas in Communications*, vol. 18, no. 10, pp. 2084–2093, 2000.
34. M. Iizuka, M. Sakuta, Y. Nishino, and I. Sasase, "A Scheduling Algorithm Minimizing Voids Generated by Arriving Bursts in Optical Burst Switched WDM Network," in *Proc. IEEE Globecom*, 2002, pp. 2736–2740.
35. M. Ljolje, R. Inkret, and B. Mikac, "A Comparative Analysis of Data Scheduling Algorithms in Optical Burst Switching Networks," in *Proc. Conference on Optical Network Design and Modeling*, 2005, pp. 493–500.
36. S. Q. Zheng, Y. Xiong, and H. C. Cankaya, "Hardware Design of a Channel Scheduling Algorithm for Optical Burst Switching Routers," in *Proc. SPIE*, vol. 4872, 2002, pp. 199–209.
37. J. Xu, C. Qiao, J. Li, and G. Xu, "Efficient Burst Scheduling Algorithms in Optical Burst-Switched Networks Using Geometric Techniques," *IEEE Journal on Selected Areas in Communications*, vol. 22, no. 9, pp. 1796–1811, 2004.
38. F. Farahmand, and J. P. Jue, "Look-Ahead Window Contention Resolution in Optical Burst Switched Networks," in *Proc. IEEE Workshop on High Performance Switching and Routing*, 2003, pp. 147–151.
39. S. Charcranoon, T. S. El-Bawab, J. D. Shin, and H. C. Cankaya, "Group Scheduling for Multi-Service Optical Burst Switching (OBS) Networks," *Photonic Network Communications*, vol. 11, no. 1, pp. 99-110, 2006.
40. H. Li, H. Neo, and L. J. I. Thng, "Performance of the Implementation of a PipeLine Buffering System in Optical Burst Switching Networks," in *Proc. IEEE Globecom*, 2003, pp. 2503–2507.
41. J. Li, C. Qiao, and Y. Chen, "Maximizing Throughput for Optical Burst Switching Networks," in *Proc. IEEE Infocom*, 2004, pp. 1853–1863.
42. N. Barakat, and E. H. Sargent, "Separating Resource Reservations from Service Requests to Improve the Performance of Optical Burst Switching Networks," *IEEE Journal on Selected Areas in Communications*, vol. 24, no. 4, pp. 95–107, 2006.
43. S. K. Tan, G. Mohan, and K. C. Chua, "Algorithms for Burst Rescheduling in WDM Optical Burst Switching Networks," *Computer Networks*, vol. 41, no. 1, pp. 41–55, 2003.

44. S. K. Tan, G. Mohan, and K. C. Chua, "Burst Rescheduling with Wavelength and Last-Hop FDL Reassignment in WDM Optical Burst Switching Networks," in *Proc. IEEE International Conference on Communications*, 2003, pp. 1448–1452.
45. M. H. Phung, K. C. Chua, G. Mohan, M. Motani, T. C. Wong, and P. Y. Kong, "On Ordered Scheduling for Optical Burst Switching," *Computer Networks*, vol. 48, no. 6, pp. 891–909, 2005.
46. R. Bhagwan and B. Lin, "Fast and Scalable Priority Queue Architecture for High-Speed Network Switches," in *Proc. IEEE Infocom*, 2000, pp. 538–547.
47. A. Ioannou and M. Katevenis, "Pipelined Heap (Priority Queue) Management for Advanced Scheduling in High-Speed Networks," in *Proc. IEEE International Conference on Communications*, 2001, pp. 2043–2047.

3

RELATIVE QOS DIFFERENTIATION

QoS provisioning in OBS networks can be generally classified as relative QoS provisioning and absolute QoS provisioning. With relative QoS provisioning, some traffic classes will have relatively better QoS than other classes without specifying any quantitative guarantees. Absolute QoS provisioning on the other hand provides for specific quantitative guarantees to each traffic class. Relative QoS differentiation can be applied when it is difficult to provide quantitative guarantees in complicated network scenarios. This chapter will discuss relative QoS differentiation provisioning among multiple traffic classes in OBS networks. Various mechanisms including offset time-based, burst segmentation, preemption based, header packet scheduling, proportional QoS differentiation, buffer allocation based scheme, and burst scheduling based scheme will be discussed.

3.1 Offset Time-Based Mechanisms

Different offset times of bursts from different flows affect significantly the QoS experienced by the flows since the larger the offset time for a burst, the higher its chance of successfully reserving a resource by virtue of it being able to reserve earlier. In [1, 2], an offset time-based scheme is proposed to provide relative QoS differentiation for multiple traffic classes. In the proposed offset time-based scheme, an extra offset time is set for a traffic class of higher priority.

3.1.1 Class isolation

A traffic class with a higher priority will be isolated from the effects of burst arrival from a traffic class with a lower priority. In OBS, each burst has a base offset time between the arrival of the header packet and the corresponding data burst at each node. The base offset time is usually set at $H\delta$, where H is the number of hops traversed by the burst and δ is the header packet processing time at each hop. To provide priority, additional offset times are defined for traffic classes with higher priorities. For two traffic class, let t_a^i denote the arrival time for the header packet for traffic class i, where $i = 1, 2$. Let t_o^i denote the offset time for traffic class i, where $i = 1, 2$. Let t_s^i denote the arrival time for the data burst for traffic class i, $i = 1, 2$. Then, $t_s^i = t_a^i + t_o^i$, $i = 1, 2$. If the header packet of class 1 (lower priority) arrives earlier than that of class 2, the header packet of class 1 will successfully reserve its required bandwidth. If the header packet for class 1 arrives later than that of class 2, the header packet of class 1 will still have the chance to reserve its required bandwidth if $t_s^1 > t_s^2 + b_l^2$, where b_l^2 is the burst length of class 2, which means that $t_a^1 + t_o^1 > t_a^2 + t_o^2 + b_l^2$ since $t_s^i = t_a^i + t_o^i$, $i = 1, 2$. Therefore, the difference between the offset times of class 1 and class 2 should meet the following condition to make sure the header packet of class 1 (lower priority) still can reserve the required bandwidth:

$$t_o^1 - t_o^2 > t_a^2 - t_a^1 + b_l^2.$$

3.1.2 Loss probability analysis under 100% class isolation

In [2], assuming 100% class isolation, the lower and upper bounds of loss probabilities are analyzed for an OBS node with multiple traffic classes with different offset times. To facilitate the analysis, the lower and upper bounds on the loss probabilities for a classless OBS node are investigated first. Let pb_l and pb_u denote the lower and upper bounds on the loss probability for a classless OBS node, respectively. The upper bound is reached when there is no FDL buffer. Therefore, the upper bound can be determined by the Erlang B formula $(M/M/k/k)$

$$pb_u(k,\rho) = \frac{\rho^k/k!}{\sum_{m=0}^{k} \rho^m/m!} \tag{3.1}$$

where ρ is the total traffic intensity and k is the number of wavelengths in the output link of the OBS node.

The lower bound is reached when the OBS node has FDL buffers. This can be approximately modelled by a $M/M/k/D$ queue, where $D = k + kN$ is the maximum number of simultaneous bursts in the OBS node, and N is the number of FDLs attached to each wavelength channel. Therefore, the lower bound can be obtained as follows:

$$pb_l(k,\rho,D) = \frac{\rho^D p_0}{k^{D-k}k!} \tag{3.2}$$

where

$$p_0 = \left(\sum_{n=0}^{k-1} \frac{\rho^n}{n!} + \sum_{n-k}^{D} \frac{\rho^n}{k^{n-k}k!}\right)^{-1}$$

Let pb_l^i and pb_u^i denote the lower and upper bounds on the loss probability of class i traffic in an OBS node with multiple traffic classes, respectively. For the analysis, it is assumed that the conservation law holds when the traffic intensity is high. This means that the overall loss probability of the network remains the same no matter how many traffic classes there are and how these traffic classes interact if the total traffic load remains the same. In other words, the overall loss probability of the classless OBS node will be approximately the same as that of the prioritized multi-class OBS node. With the isolation assumption and conservation law, the lower and upper bounds on the loss probability of class i traffic can be derived as follows.

The loss probability for the class n traffic is considered first since it has the highest priority of all traffic classes. Due to the class isolation assumption, its lower and upper bound are independent of all other traffic classes and are determined only by its own traffic intensity. Then, the upper bound and lower bound for the class n traffic can be obtained as follows according to (3.1) and (3.2), respectively:

$$pb_u^n(k, \rho_n) = \frac{\rho_n^k / k!}{\sum_{m=0}^{k} \rho_n^m / m!} \tag{3.3}$$

$$pb_l^n(k, \rho_n, D) = \frac{\rho_n^D p_0}{k^{D-k} k!} \tag{3.4}$$

where

$$p_0 = \left(\sum_{n=0}^{k-1} \frac{\rho_n^n}{n!} + \sum_{n-k}^{D} \frac{\rho_n^n}{k^{n-k} k!} \right)^{-1}.$$

For class $n-1$, due to the class isolation assumption, the average loss probability of classes n and $n-1$ is independent of all other traffic from class $n-2$ to class 1 and is determined only by the traffic intensity of classes n and $n-1$. According to the conservation law, the average loss probability of classes $n-1$ and n can be obtained as follows according to (3.1) and (3.2), respectively:

$$pb_u^{n,n-1}(k, \rho_{n,n-1}) = \frac{\rho_{n,n-1}^k / k!}{\sum_{m=0}^{k} \rho_{n,n-1}^m / m!} \tag{3.5}$$

where $\rho_{n,n-1} = \rho_n + \rho_{n-1}$.

$$pb_l^{n,n-1}(k, \rho_{n,n-1}, D) = \frac{\rho_{n,n-1}^D p_0}{k^{D-k} k!} \tag{3.6}$$

where

$$p_0 = \left(\sum_{n=0}^{k-1} \frac{\rho_{n,n-1}^n}{n!} + \sum_{n-k}^{D} \frac{\rho_{n,n-1}^n}{k^{n-k} k!} \right)^{-1}.$$

On the other hand, the average loss probability of classes $n-1$ and n can be obtained by the weighted sum of loss probabilities of classes $n-1$ and n as follows.

$$pb_l^{n,n-1} = \frac{\rho_n}{\rho_{n,n-1}} pb_l^n + \frac{\rho_{n-1}}{\rho_{n,n-1}} pb_l^{n-1} \tag{3.7}$$

and

$$pb_u^{n,n-1} = \frac{\rho_n}{\rho_{n,n-1}} pb_u^n + \frac{\rho_{n-1}}{\rho_{n,n-1}} pb_u^{n-1} \tag{3.8}$$

Then, the lower bound on loss probability of class $n-1$ can be obtained by

$$pb_l^{n-1} = \frac{pb_l^{n,n-1} - \frac{\rho_n}{\rho_{n,n-1}} pb_l^n}{\frac{\rho_{n-1}}{\rho_{n,n-1}}}$$

where $pb_l^{n,n-1}$ and pb_l^n can be obtained by (3.6) and (3.4), respectively. The upper bound on loss probability of class $n-1$ can be obtained as

$$pb_u^{n-1} = \frac{pb_u^{n,n-1} - \frac{\rho_n}{\rho_{n,n-1}} pb_u^n}{\frac{\rho_{n-1}}{\rho_{n,n-1}}}$$

where $pb_u^{n,n-1}$ and pb_u^n can be obtained from (3.5) and (3.3), respectively.

The above procedure can be repeated to obtain the lower and upper bounds on the loss probabilities for all other traffic classes from classes $n-2$ to class 1.

3.1.3 Discussion

Although this offset time based QoS provisioning scheme can give some quantitative guarantee on the loss probability in terms of lower and upper bounds for different traffic classes, it is based on the assumption that there is 100% class isolation. Without 100% class isolation, the loss probability of traffic with higher priorities will still be affected by traffic with lower priorities, which will invalidate the above analysis. It has been shown in [2] that the difference between the offset time of class i and class $i-1$ needs to be 5 times the maximum burst length size of class $i-1$, in order to achieve 99% class isolation under certain conditions. This additional offset time will lead to unacceptable end-to-end delays for traffic classes with higher priorities. In another words, to have acceptable end-to-end delays, 100% class isolation is not possible for this offset time-based scheme. Therefore, this offset time based scheme can only be considered as a relative QoS provisioning scheme.

It is reported in [3] that the offset time-based QoS algorithm does not work well when TCP is used over OBS networks. The QoS performance metric used is TCP throughput which is affected by burst loss in OBS networks. Throughputs of multiple TCP flows are measured under different values of round trip time (RTT).

Data bursts from the TCP flow with the highest priority have the longest offset time. It is shown that when the RTT is short, TCP flows with the highest priority will use most of the bandwidth and frequently preempt data bursts from other flows with lower priorities. In addition, all flows with lower priorities have almost the same throughput regardless of the differences in their offset times, which is not desirable. Only when the RTT is large enough, the flow with the highest priority will not use most of the bandwidth, allowing flows with different priorities to perform differently.

3.2 Burst Segmentation

To avoid extra delay incurred by the offset-based QoS differentiation scheme, some schemes [4, 5, 6, 7, 8, 11], have been proposed to achieve QoS differentiation by burst segmentation. The basic idea is to split one burst into different segments. When contention occurs at a core node, the overlapping segments of the contending bursts will be dropped. For ease of exposition, consider the case where there are only two bursts involved in a contention. The burst which arrives at the node first is called the original burst, and the burst which arrives later is called the contending burst. There are two choices to select from which burst the segments will be dropped as shown in Figure 3.1. One is to drop the segmented head of the contending burst. The other is to drop the segmented tail of the original burst. If the dropped segments will be retransmitted, the advantage of tail dropping is that it can help to reduce the data reordering problem.

To further reduce data loss, this approach is often combined with deflection routing such that the overlapping segment can be deflected if there are free ports available. Obviously, the burst with higher priority will win the contention. In case that the two bursts have the same priority, the longer burst will win the contention to improve the overall throughput. In particular, the following polices can be used for contention resolution according to the burst priorities and lengths.

Segment header	Segment payload	Segment header	Segment payload	◦ ◦ ◦	Segment header	Segment payload

(a) Segment structure of a burst

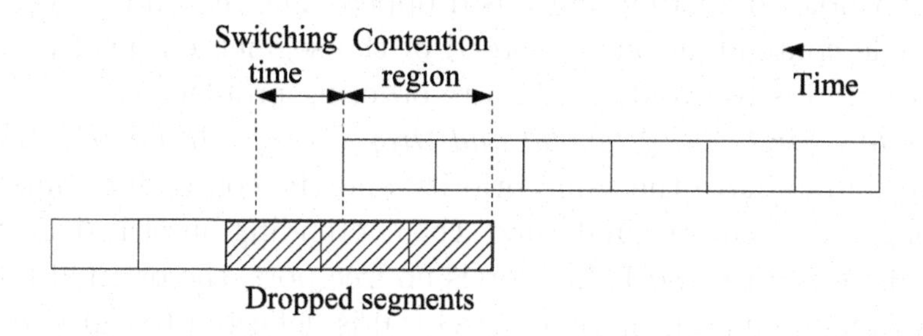

(b) Head dropping

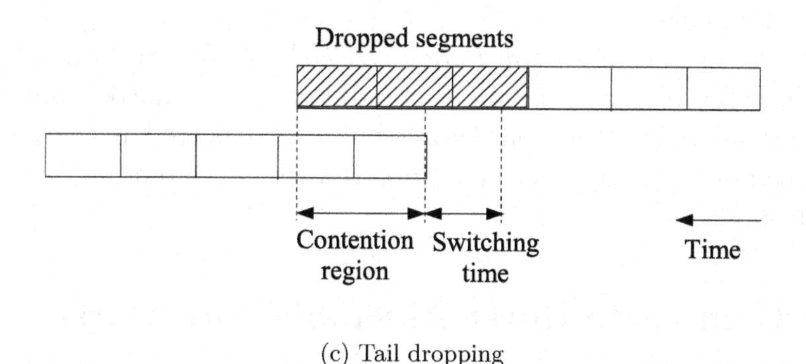

(c) Tail dropping

Fig. 3.1. Burst segmentation approaches

- *Tail Segment First and Deflect Policy (TSFDP):* If the contending burst has higher priority over the original burst, the original burst is segmented and the segmented tail is deflected to one free port. If there is no free port available, the segmented tail of the original burst is dropped.
- *Deflect First, Head Segment and Drop Policy (DFHSDP):* If the contending burst has lower priority over the original burst, the contending burst is deflected first if there are available ports for the whole contending burst. If no free port for the whole contending burst is available, the policy checks whether there

is a free port for the segmented head of the contending burst. If there is such a free port, the segmented head of the contending burst is deflected. The remaining part of the contending burst is transmitted. If there is no free port for the segmented head, the whole contending burst is dropped. The reason to drop the whole contending burst instead of the segmented head in this situation is to avoid the data re-ordering problem.

- *Deflect First, Tail Segment and Drop Policy (DFTSDP):* If the two bursts have the same priority and the contending burst is longer than the original burst, the contending burst is deflected if there is a free port. If there is no free port, the original burst is segmented and the segmented tail is deflected to a free port. If there is no free port for the segmented tail, the segmented tail will be dropped. The rest of the contending burst can be transmitted.
- *Deflect First, and Drop Policy (DFDP):* If the two bursts have the same priority and the contending burst is shorter than the original burst, the contending burst is deflected if there is a free port. If there is no free port, the whole contending burst is dropped.

3.3 Composite Burst Assembly with Burst Segmentation

In [4, 5], relative differentiated QoS schemes using composite burst assembly and burst segmentation techniques have been proposed. The basic idea is to assemble a burst using traffic from multiple traffic classes with different priorities in such a way that the data with high priority is put towards the head of each burst and the data with low priority is put towards the end of each burst. In case of a contention, the data at the end is more likely to be dropped than the data at the head of each burst with burst segmentation. The number of traffic classes and the number of priorities supported by the network will have impact on the composite burst assembly algorithms. The task of the composite burst assembly algorithm is to determine how to select packets from different traffic classes for a composite burst and how to map the composite burst to different priorities supported by the network. Let N and M denote the number of traffic classes and priorities supported by the network, respectively. Particularly, according to the relation of N and M, there are two approaches for composite burst assembly to achieve differentiated QoS in terms of loss probability.

N = M

Since the number of priorities supported by the network is M, the number of the composite burst type K equals M. In this approach, the composite burst type k is assembled with packets from traffic classes k and $k+1$ if $k \leq M-1$. The burst type M is assembled only using packets from traffic class M. The assembler will maintain separate queues for incoming packets from the different traffic classes. When it is time to assemble a composite burst with type k, the packets in the queue for traffic classes k and $k+1$ are placed into the composite burst in decreasing order of priorities, such that packets from traffic class $k+1$ are put ahead of packets from traffic class k. Both threshold-based and timeout-based burst assembly algorithms can be used here to decide when to output the composite burst. The priority of the composite burst is set as k. In case of contention, the segmented tail is more likely to be deflected or dropped depending on the availability of free ports such that packets from traffic class $k+1$ will have lower loss probability than those from traffic class k.

N > M

The number of burst type K is set as M, and the composite burst type k is assembled with packets from $\left\{\frac{(k-1)N}{M}+1, \cdots, \frac{kN}{M}\right\}$. For example, $M = 2$ and $N = 4$. Then, the number of composite burst type is set as 2. The composite burst type 1 will be assembled with packets from traffic classes 1 and 2, and the composite burst type 1 will be assembled with packets from traffic classes 3 and 4.

Although the segmentation-based approach and composite burst assembly approach can achieve differentiated QoS without setting extra offset time for each burst, these approaches increase the complexities of the burst assembler in the edge nodes and the burst scheduler in the core nodes. To solve this problem, the following probabilistic preemption-based mechanism is proposed in [9, 10] to achieve differentiated QoS with reduced complexity.

3.4 Probabilistic Preemption-Based Mechanism

Probabilistic preemption provides relative loss rate differentiation by setting different preemption probabilities for the multiple traffic classes. The basic idea is to allow bursts with higher priorities to preempt bursts with lower priorities in case of contention. To avoid starving bursts with lower priorities and achieve a desirable burst loss ratio among the different traffic classes, bursts with higher priorities preempt bursts with lower probabilities with a certain preemption probability p. By changing the preemption probability p, a flexible burst loss ratio among different traffic classes can be obtained. The details of the analysis on the probabilistic preemption can be found in [9, 10]. In [10], burst segmentation is also used in case of contention.

3.5 Header Packet Scheduling

All the QoS provisioning schemes discussed above make decision on burst scheduling immediately upon receiving the header packet. For example, in the preemption and segmentation-based schemes, bursts with lower priority are preempted or segmented immediately when there is contention for the bandwidth resource requested by multiple header packets. However, the header packet for those preempted or segmented bursts have already been sent to downstream nodes during the time the preemption or segmentation occurs. The header packets will unnecessarily reserve bandwidth at all downstream nodes, leading to non-negligible bandwidth wastage at the downstream nodes. In [12, 13], algorithms are proposed to avoid burst preemption by deferring bandwidth allocation. The basic approach is to buffer the header packet at each core node for a certain time. All header packets arrive in a certain time window are buffered into several queues according to their priorities. At the end of each time window, the bandwidth resource is allocated. The priority will be given to bursts with higher priorities, such that the desired differentiated burst blocking probability can be achieved. The benefit of this approach is to avoid the wastage of bandwidth reserved by preempted burst of lower priorities. But the disadvantage is that each header packet experiences an extra delay at each hop. Therefore, the offset time for each burst needs to be increased accordingly to make sure that the bandwidth reservation process completes before each data burst arrives at a node.

In [12], two traffic classes are considered, namely real-time traffic and non-real-time traffic. Therefore, each core node maintains two queues for the header packets from the real-time traffic and the non-real-time traffic. The window size is set as Δt. All header packets that arrive within the k-th time window $[t_k, t_k + \Delta t]$ are buffered into the two separate queues until $t_k + \Delta t$. At the end of the k-th time window, i.e. $t_k + \Delta t$, bandwidth is reserved for all bursts in the real-time queue first. Note that bandwidth needs to be reserved for a burst at the given time specified by the header packet. If there is no available bandwidth at the given time, the burst is dropped. Any remaining bandwidth after the real-time queue is served is reserved for bursts in the non-real-time queue. The bandwidth allocation discipline within each queue is first come and first served. This algorithm is greedy since all bandwidth will be reserved by bursts of high priority first.

3.6 Proportional QoS Differentiation

With proportional QoS differentiation, the goal is to provide controllable differentiated QoS such that a given QoS metric of each traffic class is proportional to some differentiation factors. Thus, each traffic class i is associated with a differentiation factor s_i. If q_i denotes a given QoS performance metric for traffic class i, then relative to another traffic class j,

$$\frac{q_i}{q_j} = \frac{s_i}{s_j} \tag{3.9}$$

1. Initialize all counters for all classes;
2. Upon arrival of one burst from class i,
   IF there is no wavelength or FDL available for this burst,
   The incoming burst is dropped, and $L_i = L_i + 1$,
   and $A_i = A_i + 1$;
   ELSE IF $P_i/s_i \leq \sum_{i=0}^{N-1} P_i/\sum_{i=0}^{N-1} s_i$,
   The incoming burst is intentionally dropped, and
   $L_i = L_i + 1$, and $A_i = A_i + 1$;
   ELSE
   The resource for the incoming burst is successfully
   reserved, and $A_i = A_i + 1$;
   END IF
3. P_i is updated.
4. **IF** $Max\left\{\left|\frac{P_i}{\sum_{i=0}^{N-1} P_i} - \frac{s_i}{\sum_{i=0}^{N-1} s_i}\right|, i = 0, \cdots, N-1\right\} \leq \varepsilon$
   Go to Step 1;
   ELSE
   Go to Step 2;
   END IF

Fig. 3.2. Proportional burst loss algorithm

3.6.1 Proportional burst loss provisioning

In this case, the QoS metric is burst loss. Each node maintains a counter to record the number of burst lost for each traffic class. A burst of low priority will be intentionally dropped if the proportionality relation (3.9) does not hold. Let L_i denote the number of burst lost for traffic class i. Let A_i denote the number of burst arrivals for class i. Let $P_i = \frac{L_i}{A_i}$ denote the loss rate for class i. The details of the algorithm are shown in Figure 3.2.

3.6.2 Proportional packet average delay provisioning

In this case, the QoS metric is average delay experienced by a packet during burst assembly. The Waiting Time Priority (WTP) scheduling algorithm [15] used in IP networks is extended to provide proportional packet average delay in OBS networks. At an edge node, each incoming packet is put into the queue for its traffic class. In this algorithm, a burst is assembled from packets from only one queue. The modified WTP scheduler runs at the burst assembler and determines from which queue the packets are used to assemble a burst. There is also a token generator inside the WTP scheduler and the token generation process is a Poisson process. When a token is generated, the packets from the selected queue are used to assemble a burst for transmission. The WTP scheduler selects the queue with the smallest $\frac{w_i}{s_i}$, where w_i is the waiting time of the packet at the head of queue i, such that proportional packet average delay can be achieved. A burst size threshold can be set to limit the number of packets assembled in a burst.

3.7 Buffer Allocation Based Schemes

Relative QoS can be provided through buffer allocation in OBS networks both at edge nodes and core nodes (FDLs).

3.7.1 Buffer allocation in edge nodes

In [16], a buffer allocation scheme has been proposed to provide service differentiation of the edge nodes. There are n service classes in an OBS network. A time limit d_n is set for service class n, which sets the maximum time that a data burst from class n is allowed to be stored in the buffer at an edge node. All incoming data bursts are inserted into a FIFO queue. When a class n burst that arrives at time t reaches the head of the FIFO queue, the scheduler searches for a wavelength that is free for the burst. If no free wavelength is found until time $t+d_n$, the burst is dropped. By providing different time limits to different service classes, differentiated loss probabilities can be achieved in the edge node.

Another buffer allocation-based algorithm has been proposed in [17]. This algorithm is inspired by the random early dropping (RED) scheme which is used in traditional IP-based networks for congestion control. In RED, incoming packets to a node are dropped with different probabilities proactively based on the average queue length at the node. In the RED-like algorithm in [17], incoming bursts are similarly proactively dropped when the queue length of the edge node exceeds a given threshold. Each incoming burst is proactively dropped with a probability determined by the priority of the service class of the incoming burst. Thus, by setting different dropping probabilities for different service classes, service differentiation can be achieved at an edge node.

3.7.2 FDL allocation in core nodes

At a core node, FDLs can also be allocated to different service classes for providing differentiated QoS. In [16], each service class is allocated a quota on the maximum number of FDLs to use to delay data bursts. Each core node monitors the number of FDLs in use for each service class. A data burst from a service class is dropped if there is no FDL available within its allocated quota. The service class with the highest priority has the largest quota of FDLs.

3.8 Burst Scheduling Based Scheme

3.8.1 Bandwidth usage profile-based algorithms

Differentiated QoS can also be provided by allocating different amount of bandwidth at the core nodes to different service classes. In [18], a service class i is allocated a predefined bandwidth usage limit p_i, which is the fraction of bandwidth usage that the class is allowed to use. Therefore, $\sum_{i=1}^{N} p_i = 1$, where N is the number of service classes. Each core node monitors and records the usage of bandwidth for each service class. The usage of bandwidth for class i, ρ_i is obtained according to the formula:

$$\rho_i = \frac{\sum_{j=1}^{n_i} l_j}{\sum_{i=1}^{N} \sum_{k=1}^{n_i} l_k}$$

where l_j is the burst size of data bursts from class j, n_j is the number of scheduled bursts from class j within a certain measurement time window. Therefore, $\sum_{j=1}^{n_i} l_j$ is the bandwidth usage of class j during the measurement time window. $\sum_{j=1}^{N} \sum_{k=1}^{n_j} l_k$ is the total bandwidth usage for all the service classes during the measurement time window. Class i is said to be in profile if $\rho_i \leq p_i$; otherwise, it is said to be out of profile.

When a node receives a header packet from class i and there is no available wavelength for the corresponding data burst, the node determines whether class i is in profile or not. If it is out of profile, the header packet is simply dropped without. If it is in profile, the node constructs a preemption list which consists of all the scheduled bursts which overlap with the incoming data burst. Only overlapping bursts from out of profile classes are put into the preemption list. The burst from the service class with the lowest priority in the preemption list is then preempted by the incoming burst.

There are two options to preempt existing bursts. One is called partial preemption, the other is called full preemption. In partial preemption, only the overlapping part of the existing burst is preempted by the incoming burst, which is similar to burst segmentation. In full preemption, the whole existing burst is preempted.

3.8.2 A wavelength search space-based algorithm

For OBS networks without QoS differentiation, each incoming header packet is allowed to search all wavelengths for available bandwidth. By limiting the wavelength search space for each header packet from different service classes, differentiated QoS can be provided. Intuitively, a service class with a larger wavelength search space will have a lower burst loss probability. In [17], a wavelength search space-based algorithm is proposed to support differentiated QoS. In this algorithm, each service class is associated with a search space which consists of two parts. The first part is the base search space which is the same for all service classes to avoid the problem of service starvation for service classes with lower priorities. The second part is the adjustable search space which is associated with the priority of each service class. The higher the priority of a service class, the larger the adjustable search space for that service class. Assume that service class i has higher priority over service class $i-1, i-2, \cdots, 1$. For a burst from service class i, the search space n_i for class i is given by

$$n_i = (1-g)\,W + giW/P \tag{3.10}$$

where $g \in [0,1]$ is a controllable parameter, W is the total number of wavelengths, and P is the total number of service classes. It can be seen from Eq. (3.10) that $(1-g)\,W$ is the base search space part which is independent of the priority of the given service class, and giW/p is the adjustable part which is dependent on the priority of the given service class. It can be easily verified that the service class with the highest priority can search all the wavelengths for available bandwidth.

By adjusting the controllable parameter g, various differentiation levels among service classes can be achieved. However, the parameter g has significant impact on the performance of this algorithm. With a larger g, the algorithm can provide more effective service differentiation. However, service classes with lower priorities will be significantly affected with a larger g. In addition, with a larger g, the search space for service classes with lower priorities will be quite small, and some bandwidth will be unnecessarily wasted if there is no enough high priority traffic. The consequence is that the overall throughput of the system will be reduced. Therefore, a proper value of g needs to be selected to provide a desirable QoS differentiation without significant degradation on the system throughput.

References

1. M. Yoo and C. Qiao, "A New Optical Burst Switching Protocol for Supporting Quality of Service," in *Proc. SPIE Conference on All-Optical Networking*, vol. 3531, 1998, pp 396–405.
2. M. Yoo, C. Qiao, and S. Dixit, "QoS Performance of Optical Burst Switching in IP-Over-WDM networks," *IEEE Journal on Selected Areas in Communications*, vol.18, pp 2062–2071, 2000.
3. M.-G. Kim, H. Jeong, J.Y. Choi, J.-H. Kim, and M. Kang , "The Impact of the Extra Offset-Time based QoS Mechanism in TCP over Optical Burst Switching Networks," in *Proc. Optical Communication Systems and Networks*, 2006.
4. V. Vokkarane, Q. Zhang, J.P. Jue, and B. Chen, "Generalized Burst Assembly and Scheduling Techniques for QoS Support to Optical Burst-Switched Networks," in *Proc. IEEE Globecom*, 2002, pp. 2747–2751.
5. V. Vokkarane and J.P. Jue, "Prioritized Burst Segmentation and Composite Burst-Assembly Techniques for QoS Support to Optical Burst-Switched Networks," *IEEE Journal Selected Areas Communications*, vol. 21, no. 7, pp. 1198–1209, 2003.
6. V. Vokkarane, J. P. Jue, and S. Sitaraman, "Burst Segmentation: An Approach for Reducing Packet Loss in Optical Burst Switched Networks," in *Proc. IEEE International Conference on Communications*, 2002, pp. 2673–2677.
7. V. Vokkarane and J. P. Jue, "Prioritized Routing and Burst Segmentation for QoS in Optical Burst- Switched Networks," in *Proc. IEEE/OSA Optical Fiber Communication Conference*, 2002, pp. 221–222.
8. V. M. Vokkarane and J. P. Jue, "Segmentation-Based Nonpreemptive Scheduling Algorithms for Optical Burst-Switched Networks," *Journal of Lightwave Technology*, vol. 23, no. 10, pp.3125-3137, 2005.
9. L. Yang, Y. Jiang, and S. Jiang, "A Probabilistic Preemptive Scheme for Providing Service Differentiation in OBS Networks," in *Proc. IEEE Globecom*, 2003, pp. 2689–2693.
10. C.W. Tan, G. Mohan, and J. C. Lui, "Achieving Proportional Loss Differentiation Using Probabilistic Preemptive Burst Segmentation in Optical Burst Switching WDM Networks," To appear in *IEEE Journal on Selected Areas in Communications*, Optical Communications and Networking Series, 2006.

11. A. Abid, F.M. Abbou, and H.T. Ewe, "Effective Implementation of Burst Segmentation Techniques in OBS Networks," *International Journal of Information Technology*, vol.3, no. 4, pp 231–236, 2006.
12. Y. Wang and B. Ramamurthy, "CPQ: A Control Packet Queuing Optical Burst Switching Protocol for Supporting QoS," in *Proc. Third Workshop on Optical Burst Switching*, 2004.
13. F. Farahmand and J. Jue, "Supporting QoS with Look-ahead Window Contention Resolution in Optical Burst Switched Networks," in *Proc. IEEE Globecom*, 2003, pp. 2699–2673.
14. Y. Chen, M. Hamdi, and D.H.K. Tsang, "Proportional QoS over OBS networks," in *Proc. IEEE Globecom*, 2001, pp. 1510–1514.
15. C. Dovrolis, D. Stiliadis, and P. Ramanathan, "Proportional Differentiated Services: Delay Differentation and Packet Scheduling," in *Proc. ACM Sigcomm*, 1999, pp. 109–120.
16. J, Phuritatkul and Y. Ji, "Resource Allocation Algorithms for Controllable Service Differentiation," *IEICE Transaction on Communication*, vol.E88-B, no. 4, pp 1424–1431, 2005.
17. B. Zhou, M. Bassiouni and G. Li, "Routing and Wavelength Assignment in Optical Networks Using Logical Link Representation and Efficient Bitwise Computation," *Journal of Photonic Network Communications*, Vol.10, No. 3, pp. 333–346, 2005.
18. W. Liao and C.H. Loi, "Providing Service Differentiation for Optical-Burst-Switched Networks," *Journal of Lightwave Technology* vol. 22, pp. 1651–1660, 2004.

4 ABSOLUTE QOS DIFFERENTIATION

The absolute Quality of Service (QoS) model in Optical Burst Switching (OBS) aims to give worst-case quantitative loss guaratees to traffic classes. For example, if a traffic class is guaranteed to experience no more than 0.1% loss rate per hop, the loss rate of 0.1% is referred to as the absolute threshold of that class. This kind of QoS guarantee calls for different QoS differentiation mechanisms than those intended for the relative QoS model in the previous chapter. A common characteristic of these absolute QoS mechanisms is that they differentiate traffic based on the classes' absolute thresholds instead of their relative priorities. That is, traffic of a class will get increasingly favourable treatment as its loss rate gets closer to the predefined threshold. In this way, the absolute thresholds of the classes will be preserved.

This chapter will discuss the various absolute QoS mechanisms proposed in the literature. They include early dropping, preemption, virtual wavelength reservation and wavelength grouping mechanisms. Some of them such as early dropping and preemption are also used with the relative QoS model. However, the dropping and preemption criteria here are different. The other mechanisms are unique to the absolute QoS model.

4.1 Early Dropping

Early dropping is first proposed in [1] to implement proportional QoS differentiation in OBS, which is described in the previous chapter. In this section, its use in absolute QoS differentiation as proposed in [2] will be discussed.

4.1.1 Overview

In the early dropping scheme, a node monitors the local QoS performance of all traffic classes against their required thresholds. When the performance of a particular class $j (1 \leq j \leq N)$ drops below the required level, the node selects another class i that has its QoS performance above the required level for dropping. Header packets of class i will be dropped before they reach the scheduler. Consequently, the offered load to the node is reduced and performance of other classes, including class j, will improve.

To decide which class whose header packets are to be dropped early, the node assigns class priorities based on how stringent their loss thresholds are. It then computes an *early dropping probability* $p_{C_i}^{ED}$ for each class i based on the monitored loss probability and the acceptable loss threshold of the next higher priority class. An *early dropping flag*, e_i, is associated with each class i. e_i is determined by generating a random number between 0 and 1. If the number is less than $p_{C_i}^{ED}$, e_i is set to 1. Otherwise, it is set to 0. Hence, e_i is 1 with probability $p_{C_i}^{ED}$ and is 0 with probability $(1 - p_{C_i}^{ED})$. Suppose class priorities are set such that one with a higher index has a higher priority. An *early dropping vector*, ED_i, is generated for the arriving class i burst, where $ED_i = \{e_i, e_{i+1}, \ldots, e_{N-1}\}$. The class i header packet is dropped if $e_i \vee e_{i+1} \vee \cdots \vee e_{N-1} = 1$. In other words, it is dropped early if any of the early dropping flags of its class or higher priority classes is set. Thus, the arriving class i burst is dropped with probability $(1 - \prod_{j=i}^{N-1}(1 - p_{C_j}^{ED}))$. Note that an element e_N for class N is not included in the formula because class N has the highest priority.

Although the early dropping approach provides adequate differentiation among classes, it is not very efficient. This is because it may unnecessarily drop low priority bursts that otherwise would not contend with any other bursts. This leads to under-utilization of wavelengths. Preemption, which will be described later, overcomes this drawback.

4.1.2 Calculation of the early dropping probability

A key parameter of the early dropping scheme is the early dropping probability $p_{C_i}^{ED}$ for each class i. Two methods to calculate $p_{C_i}^{ED}$ are proposed in [2], namely Early Drop by Threshold (EDT) and Early Drop by Span (EDS).

In the EDT method, all bursts of class i are early dropped when the loss probability of the next higher priority class $(i+1)$, $p_{C_{i+1}}$, reaches the acceptable loss threshold $P_{C_{i+1}}^{MAX}$. The early dropping probability $p_{C_i}^{ED}$ of class i is given by

$$p_{C_i}^{ED} = \begin{cases} 0, & p_{C_{i+1}} < P_{C_{i+1}}^{MAX} \\ 1, & p_{C_{i+1}} \geq P_{C_{i+1}}^{MAX} \end{cases} \tag{4.1}$$

where $i \geq 1$.

Since this method has only one single trigger point, bursts of each class with lower priority than class $(i+1)$ suffer from high loss probability when $p_{C_{i+1}}$ exceeds $P_{C_{i+1}}^{MAX}$.

To avoid the above negative side effect of the EDT method, the EDS method linearly increases p_{C+i}^{ED} as a function of $p_{C_{i+1}}$ over a span of acceptable loss probabilities $\delta_{C_{i+1}}$. The EDS algorithm is triggered when the loss probability of class $i+1$, $p_{C_{i+1}}$, is higher than $P_{C_{i+1}}^{MIN} = P_{C_{i+1}}^{MAX} - \delta_{C_{i+1}}$. Thus, the early dropping probability $p_{C_i}^{ED}$ of class i is given by

$$p_{C_i}^{ED} = \begin{cases} 0, & p_{C_{i+1}} < P_{C_{i+1}}^{MIN} \\ \frac{p_{C_{i+1}} - P_{C_{i+1}}^{MIN}}{\delta_{C_{i+1}}}, & P_{C_{i+1}}^{MIN} \leq p_{C_{i+1}} < P_{C_{i+1}}^{MAX} \\ 1, & p_{C_{i+1}} \geq P_{C_{i+1}}^{MAX} \end{cases} \tag{4.2}$$

where $i \geq 1$.

4.2 Wavelength Grouping

Wavelength grouping is an absolute QoS differentiation mechanism proposed in [2]. In this scheme, each traffic class i is provisioned a minimum number of wavelengths W_{C_i}. The Erlang B formula is used to determine W_{C_i} based on the maximum offered load L_{C_i}

and the maximum acceptable loss threshold $P_{C_i}^{MAX}$ of class i as follows

$$\frac{\frac{L_{C_i}^{W_{C_i}}}{W_{C_i}!}}{\sum_{n=0}^{W_{C_i}} \frac{L_{C_i}^n}{n!}} \leq P_{C_i}^{MAX}. \tag{4.3}$$

If the total number of required wavelengths is larger than the number of wavelengths on a link then the requirements of some classes cannot be satisfied with the given link capacity. On the other hand, if wavelengths are still available after provisioning wavelengths for all guaranteed classes, the remaining wavelengths can be used for best effort traffic.

Two variants of the wavelength grouping scheme are proposed, namely, Static Wavelength Grouping (SWG) and Dynamic Wavelength Grouping (DWG). In the SWG variant, the wavelengths assigned for each traffic class i are fixed and class i bursts are only scheduled on those assigned wavelengths. In this way, the set of available wavelengths are partitioned into disjoint subsets and each subset is provisioned exclusively for one class. The disadvantage of this method is low wavelength utilization since some bursts of class i may be dropped even if wavelengths provisioned for other classes are free. The DWG variant of the wavelength grouping scheme avoids this drawback by allowing class i bursts to be scheduled on any wavelength as long as the total number of wavelengths occupied by class i bursts is less than W_{C_i}. To implement this, a node must be able to keep track of the number of wavelengths occupied by bursts of each class.

The wavelength grouping approach has the drawback of inefficient wavelength utilization. The reason is that bursts of a given class are restricted to a limited number of wavelengths. Therefore, if a class experiences a short period of high burst arrival rate, its bursts cannot be scheduled to more wavelengths even though they may be free.

Table 4.1. Integrated Scheme: Label Assignment

e_0	Class 0 label	Class 1 label
0	L1	L1
1	L0	L1

4.3 Integrating Early Dropping and Wavelength Grouping Schemes

Since both early dropping and wavelength grouping differentiation schemes result in inefficient wavelength utilization, the integrated scheme is proposed in [2] as a way to alleviate this drawback. This is a two-stage differentiation scheme. In the first stage, the Early Drop by Span (EDS) algorithm is used to classify bursts into groups based on the class of the burst and the current value of the corresponding early dropping vector. Bursts in the same group are given the same label. In the second stage, the wavelength grouping algorithm provisions a minimum number of wavelengths for each group and schedules each burst accordingly.

For simplicity, we describe the integrated scheme using a two-class example. In the first phase, as shown in Table 4.1, a burst is labeled L1 if it is either a class 1 burst or a class 0 burst with $e_0 = 0$. Otherwise, it is labeled L0. The labeled burst is sent to the scheduler, which schedules it based solely on the label. Table 4.2 gives the number of wavelengths provisioned for each group of bursts with a given label. A burst labeled L1 can be scheduled on any of the W wavelengths of the link. This allows all wavelengths to be utilized when the early dropping scheme is not triggered. On the other hand, a burst labeled L0 can only be scheduled on $W - W_{C_1}$ wavelengths where W_{C_1} is the minimum number of wavelengths provisioned for class 1 as defined in section 4.2. If SWG is used in the second stage then L0 bursts can only be scheduled on $W - W_{C_1}$ fixed wavelengths. On the other hand, if DWG, is used, L0 bursts can be scheduled on any wavelength provided that the total number of wavelengths occupied by L0 bursts does not exceed $W - W_{C_1}$. This restriction ensures that class 1 bursts are always adequately provisioned.

Table 4.2. Integrated Scheme: Wavelength provisioning

Burst label	Wavelengths provisioned
L0	$W - W_{C_1}$
L1	W

4.4 Preemption

In the preemption method, when a node receives the header of a high-priority burst and fails to schedule it, the node may drop, or preempt, a scheduled low-priority burst to make room for the high-priority one. Thus, preemption employs the same concept as intentional dropping, but is more efficient and elegant in that it drops only those necessary to schedule the high-priority burst. When preemption happens, the node may send a control message to downstream nodes to clear the reservations of the preempted burst. Alternatively, it may choose to do nothing, which leads to some inefficiency as those reserved time intervals at the downstream nodes cannot be utilised by other bursts.

Preemption is a popular QoS differentiation mechanism and is used to implement both the relative QoS model and the absolute QoS model. The key part is proper definition of the preemption policy according to the intended QoS model. The preemption policy determines in a contention which burst is the "high-priority" one, i.e., having preemption right. Apart from many proposals to use preemption for relative QoS differentiation, preemption is used in [3, 4, 5, 6] to implement absolute QoS differentiation. The proposal in [5] is designed for Optical Packet Switching (OPS). Since it is also applicable for OBS with little modification, it will be considered in the following discussion. The common feature of the preemption policy in those proposals is that bursts belonging to a class whose loss probability approaches or exceeds its threshold are able to preempt bursts from other classes.

4.4.1 Probabilistic preemption

In [5], a probabilistic form of preemption is used to implement absolute differentiation between two classes as follows. The high priority class 1 is assigned a loss probability band $(P_{1,min}, P_{1,max}$ and a preemption probability $p(0 \leq p \leq 1)$. In a contention with a low priority burst, the high priority burst has preemption right with probability p. The parameter p is adjusted to make the loss probability of the high priority class fall within the band.

The adjustment of p is done in cycles. Each cycle consists of a specific number of class 1 bursts. In cycle n, the loss probability for class 1, $P_{1,est}$, is measured. The parameter p is adjusted in the next cycle $n+1$ according to the formula below

$$p_{n+1} = \begin{cases} p_n - (1-p_n)\delta, & P_{1,est} < P_{1,min} \\ p_n + (1-p_n)\delta, & P_{1,min} \leq P_{1,est} < P_{1,max} \\ p_n + (1-p_n)\delta^{0.5}, & P_{1,est} \geq P_{1,max} \end{cases} \quad (4.4)$$

where $\delta(0 < \delta < 1)$ is the adjustment factor. Note that p is only defined within $(0 \leq p \leq 1)$. Therefore, any adjustment that will take p outside these bounds are ignored.

4.4.2 Preemption with virtual channel reservation

In [6], preemption is used in combination with a virtual channel reservation (VCR) scheme to implement absolute QoS differentiation. Suppose that a node has T wavelengths per output link and N traffic classes, which are denoted as $c_1, c_2, \ldots, c_N$ with c_N being the highest priority class. The lower the required loss threshold of a class, the higher its priority. The switch assigns each class i a parameter $k_i(0 \leq k_i \leq T)$, which is the maximum number of wavelengths that class i bursts are allowed to occupy. The parameter k_i is determined based on the loss threshold for class i using the Erlang B formula similar to equation (4.3). The preemption with virtual channel reservation algorithm is described below.

In normal operation, k_i is usually dormant. It is only applied when all wavelengths are occupied. Specifically, if a free wavelength can be found, an incoming burst will be scheduled to that wavelength regardless of its priority class. On the other hand, if a class i burst arrives and finds all wavelengths occupied, the node will count the number of wavelengths already occupied by bursts of class i. If and only if the number is less than k_i, preemption will occur in a lower priority class $j(1 \leq j \leq i-1)$; otherwise, the incoming burst is dropped. The selection of the burst to be preempted always starts from the lowest priority class and goes up. If no lower priority burst can be found, the incoming burst will be dropped.

4.4.3 Preemption with per-flow QoS guarantee capability

In the absolute QoS model, quantitative end-to-end guarantees are intended to be offered to individual flows. In order for that to happen, there are two requirements for an absolute QoS differentiation mechanism at each core node as follows.

- *Inter-class requirement:* It must ensure that as the offered load to a link increases, the distances to thresholds of all classes present at the link converge to zero. This implies that burst loss from classes that are in danger of breaching their thresholds is shifted to other classes by the differentiation scheme.
- *Intra-class requirement:* It must ensure that bursts belonging to the same class experience the same loss probability at a particular link regardless of their offsets and burst lengths. In OBS networks, it is well-known that burst lengths and offsets have significant impacts on burst loss probability. Hence, without intervention from the differentiation scheme, some flows with unfavourable burst characteristics may experience loss probabilities above the threshold even though the overall loss probability of the class is still below the threshold.

All differentiation algorithms discussed so far in this chapter can satisfy the inter-class requirement above. That is, they can guarantee the *overall* loss probability of a class to be lower than the required loss threshold. However, none of them considers and satisfies the intra-class requirement. It is well known that burst length distribution and burst offset have major influence on the loss probability experienced by a flow. As a result, a flow with unfavourable offset or burst length distribution may experience loss probability greater than the required threshold even though the overall loss probability of the class it belongs to is still below the threshold.

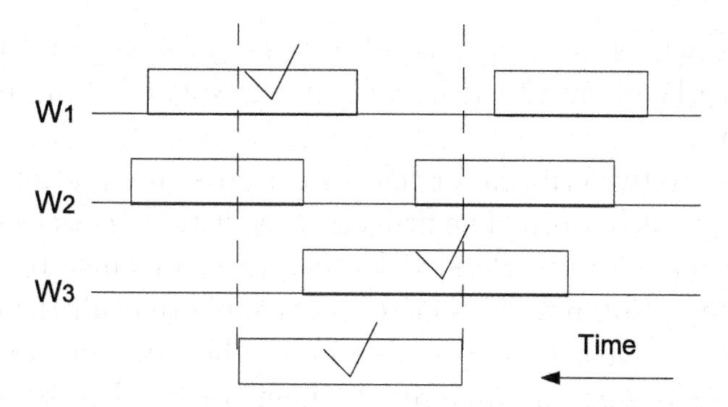

Fig. 4.1. Construction of a contention list

The preemption algorithm proposed in [3, 4] can satisfy both the inter-class and the intra-class requirements. And it achieves that while operating only at the class level. That is, it only requires a node to keep per-class information, which includes the predefined threshold, the amount of admitted traffic and the current loss probability.

The algorithm works as follows. When a header packet arrives at a node and fails to reserve an output wavelength, the node constructs a *contention list* that contains the incoming burst reservation and the scheduled burst reservations that overlap (or contend) with the incoming one. Only one scheduled reservation on each wavelength is included if its preemption helps to schedule the new reservation. This is illustrated in Figure 4.1 where only the ticked reservations among the ones overlapping with the incoming reservation on wavelengths W_1, W_2 and W_3 are included in the contention list. The node then selects one reservation from the list to drop according to some criteria described later. If the dropped reservation is a scheduled one then the incoming reservation will be scheduled in its place. That is, the incoming reservation *preempts* the scheduled reservation.

When preemption happens, a special NOTIFY packet will be immediately generated and sent on the control channel to the downstream nodes to inform them of the preemption. The downstream nodes then remove the burst reservation corresponding to the preempted burst. Although one NOTIFY packet is required for every preemption, the rate of preemption is bounded by the loss rate, which is usually kept very small. Therefore, the additional overhead by the transmission of NOTIFY packets is not significant.

There are two criteria for selecting a burst reservation from the contention list to drop. The first criterion is that the selected reservation belongs to the class with the largest distance to threshold in the contention list. This criterion ensures that all the distances to thresholds of the classes present at the node are kept equal, thereby satisfying the first requirement above. The second criterion is applied when there are more than one reservation belonging to the class with the largest distance to threshold. In that case, only one of them is selected for dropping. Let the length of the ith reservation be $l_i, (1 \leq i \leq N)$, where N is the number of reservations belonging to the class with the largest distance to threshold in the contention list. The probability of it being dropped is

$$p_i^d = \frac{\frac{1}{l_i}}{\sum_{j=1}^{N} \frac{1}{l_j}}. \tag{4.5}$$

The rationale is that the probability that a reservation is involved in a contention is roughly proportional to its length, assuming Poisson burst arrivals. So p_i^d is explicitly formulated to compensate for that burst length selection effect. In addition, the selection is independent of burst offsets. That is, although a large offset burst is less likely to encounter contention when its header packet first arrives, it is as likely to be preempted as other bursts in subsequent contention with shorter offset bursts. Therefore, the second requirement is achieved.

The above description assumes that no FDL buffer is present. It can be trivially extended to work with FDL buffers by repeating the preemption procedure for each FDL and the new reservation interval.

4.4.4 Analysis

In this section, the overall loss probability for the preemptive differentiation scheme is derived. Both the lower and upper bounds and an approximate formula for the loss probability are derived.

Depending on the application's requirement, one can choose the most suitable formula to use.

The following assumptions are used in the analysis. Firstly, for the sake of tractability, only one QoS class is assumed to be active, i.e., having traffic. The simulation results in Figure 4.3 indicate that the results obtained are also applicable to the case with multiple traffic classes. Secondly, burst arrivals follow a Poisson process with mean rate λ. This is justified by the fact that a link in a core network usually has a large number of traffic flows and the aggregation of a large number of independent and identically distributed point processes results in a Poisson point process. Thirdly, the incoming traffic consists of a number of traffic components with the ith component having a constant burst length $1/\mu_i$ and arrival rate λ_i. This assumption results from the fact that size-triggered burst assembly is a popular method to assemble bursts. This method produces burst lengths with a very narrow dynamic range, which can be considered constant. Finally, no FDL buffer is assumed and the offset difference among incoming bursts is minimal.

The lower bound on loss probability is easily derived by observing that preemption itself does not change the total number of lost bursts in the system. Thus, it is determined using Erlang's loss formula for an $M|G|k|k$ queueing model as follows:

$$P_l = B(k, \rho) = \frac{\frac{r^k}{k!}}{\sum_{i=0}^{k} \frac{r^i}{i!}}, \tag{4.6}$$

where k is the number of wavelengths per output link; ρ is the total offered load and $r = k\rho$.

Although preemption does not directly affect the number of lost bursts, it affects the probability that a burst whose header packet arrives later is successfully scheduled. Depending on the reservation intervals of later bursts, the preemption may have detrimental or beneficial effects. Consider a preemption scenario as illustrated in Figure 4.2 where burst 1 is preempted by burst 2. Let bursts 3 and 4 be two bursts whose header packets arrive after the preemption. For burst 3, the preemption is detrimental because had there been no preemption, burst 3 would be successfully scheduled. On

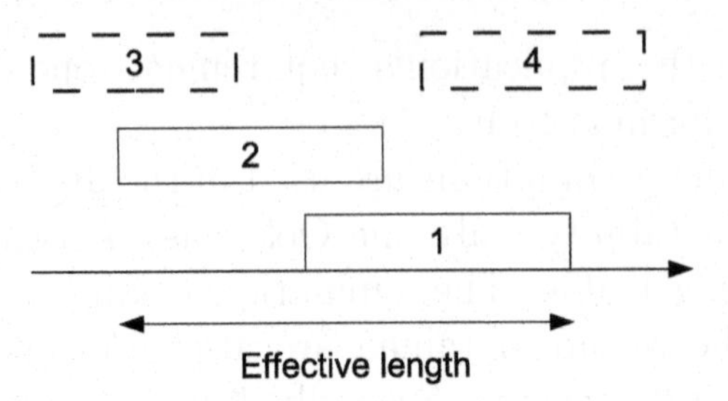

Fig. 4.2. Example of a preemption scenario.

the other hand, the preemption is beneficial to burst 4. However, for that to happen, burst 4 has to have a considerably shorter offset than other bursts, which is unlikely due to the assumption that the offset difference among bursts is minimal. For other preemption scenarios, it can also be demonstrated that a considerable offset difference is required for a preemption to have beneficial effects. Therefore, it can be argued that preemption generally worsens the schedulability of later bursts.

To quantify that effect, it is observed that from the perspective of burst 3, the preemption is equivalent to dropping burst 2 and extending the effective length of burst 1 as in Figure 4.2. Therefore, it increases the time that the system spends with all k wavelengths occupied. The upper bound on burst loss probability is derived by assuming that the loss probability is also increased by the same proportion. Denote

$$\delta = \frac{(1/\mu' - 1/\mu)}{1/\mu} = \frac{\mu}{\mu'} - 1$$

where $1/\mu'$ is the new effective length and $1/\mu$ is the actual length of the preempted burst. The upper bound on loss probability is then given as

$$P_u = \begin{cases} (1 + \delta\rho)B(k, \rho) & \text{if } B(k, \rho) < \frac{1}{1+\delta\rho} \\ 1 & \text{otherwise} \end{cases} . \tag{4.7}$$

An approximate formula for the loss probability can be derived based on (4.7) by observing that the increase in effective length of a preempted burst will increase the overall loss probability only if another incoming burst contends with it again during the extended duration. The probability that this does not happen is

$$p = \sum_{i=0}^{\infty} \frac{e^{-\delta r}(\delta r)^i}{i!} \left(\frac{k}{k+1}\right)^i = e^{-\frac{\delta r}{k+1}}. \tag{4.8}$$

From (4.7) and (4.8), the loss probability is given as

$$P = P_u - e^{-\frac{\delta r}{k+1}} \delta \rho B(k, \rho). \tag{4.9}$$

To derive δ, suppose the incoming traffic has N_c traffic components with N_c different burst lengths. Let a and b denote the component indices of the incoming burst and the preempted burst, respectively. The probability of a particular combination (a, b) is given by the formula

$$P(a,b) = \frac{\lambda_a}{\sum_{i=1}^{N_c} \lambda_i} \cdot \frac{\rho_b}{\sum_{j=1}^{N_c} \rho_j} \cdot \frac{\mu_b}{\sum_{k=1}^{N_c} \mu_k} \quad (1 \le a, b \le N_c). \tag{4.10}$$

The first and second factors are the probabilities that an incoming burst and a scheduled burst belong to components a and b, respectively. The third factor accounts for the length selective mechanism of the preemption scheme. For a preemption situation (a,b), the effective length is increased by $\frac{1}{\mu_a} - \frac{1}{2\mu_b}$. Therefore, it follows that

$$\delta = \sum_{a=1}^{N_c} \sum_{b=1}^{N_c} P(a,b) \left(\frac{\mu_b}{\mu_a} - \frac{1}{2}\right). \tag{4.11}$$

4.4.5 Numerical study

In this section, the proposed absolute differentiation algorithm is evaluated against the two criteria in section 4.4.3 and the analytical results obtained in section 4.4.4 are verified through simulation at the node level.

The node in the simulation has an output link with 64 data wavelengths, each having a transmission rate of 10 Gbps. It is assumed that the node has full wavelength conversion capability and no buffering. Bursts arrive at the link according to a Poisson process with rate λ. This Poisson traffic assumption is valid for core networks due to the aggregation effect of a large number of flows per link. The burst lengths are generated by a size-limited burst assembly algorithm with a size limit of 50 kB. Thus, the generated bursts have lengths between 50 kB and 51.5 kB, or between 40 μs and 41.2 μs.

The first experiment attempts to verify the accuracy of the analysis in section 4.4.4. For this purpose, the overall loss probabilities of traffic with one QoS class, traffic with seven QoS classes and the analytical value against the overall loading are plotted. In the case with seven classes, the classes are configured with thresholds ranging from $T_l = 0.0005$ to $T_h = 0.032$ and the ratio between two adjacent thresholds is $\gamma = 2$. The traffic of the highest threshold class takes up 40% of the total traffic. For each of the remaining classes, their traffic takes up 10% of the total traffic.

From Figure 4.3, it is observed that all the three loss curves match one another very well. This shows that the analysis is accurate and its assumption is valid, i.e., the traffic mix does not affect the overall loss probabilities. The reason is that in this differentiation scheme, preemption potentially happens whenever there is a contention between bursts, regardless of whether they are of different classes or of the same class. Therefore, the number of lost bursts depends only on the number of burst contentions and not the traffic mix.

In the next experiment, the loss probabilities of individual classes are plotted against the overall loading in Figure 4.4. For easy visualisation, only two QoS classes are assumed. Class 0 has a threshold of 0.005 and takes up 20% of the overall traffic. Class 1 has a threshold of 0.01. It is observed that as the loading increases, the loss probabilities of both classes approach their corresponding thresholds. This shows that the algorithm satisfies the first criterion set out in section 4.4.3. In addition, the distances to thresholds are always kept equal except when the loss probability of class 0 becomes zero. Although this feature is not required by the model, it is introduced to keep the minimum of the distances to thresh-

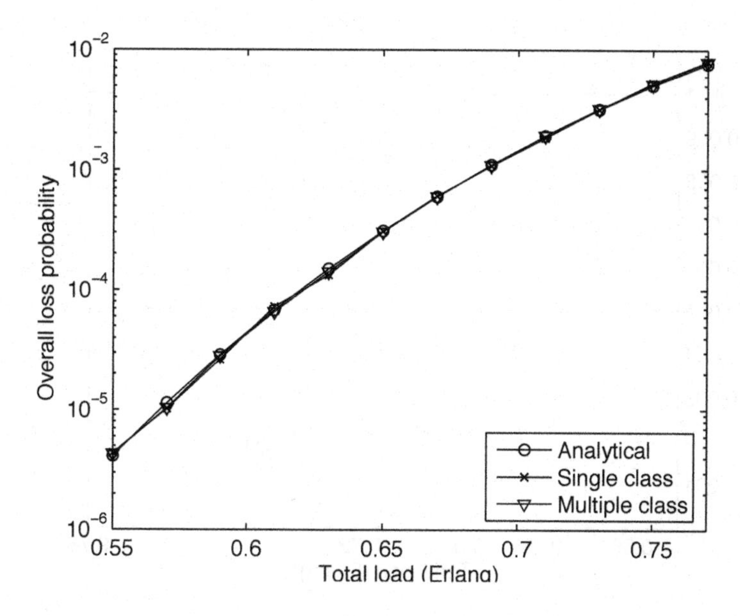

Fig. 4.3. Validation of analytical result for different number of traffic classes

olds as large as possible, thereby reducing the possibility that a sudden increase in incoming traffic could take the loss probability of a class beyond its threshold.

In Figure 4.5, the loss performance of traffic components with different traffic characteristics within a class is observed. The overall loading is 0.77 and the class configuration is the same as in the previous experiment. Each class has two traffic components in equal proportion. The plot in Figure 4.5(a) is for the situation where the traffic components have different offsets. The offset difference is 40 μs, which is approximately one burst length. In Figure 4.5(b), each class has two traffic components with different burst lengths. The size limits for the two components in the burst assembly algorithms are set at 50 kB and 100 kB, respectively. These settings would cause major differences in loss performance of the traffic components in a normal OBS system. Nevertheless, both figures show that the loss performance of different components within a class follows each other very closely despite the difference in their burst characteristics. It can be concluded that the proposed differentiation scheme can achieve uniform loss per-

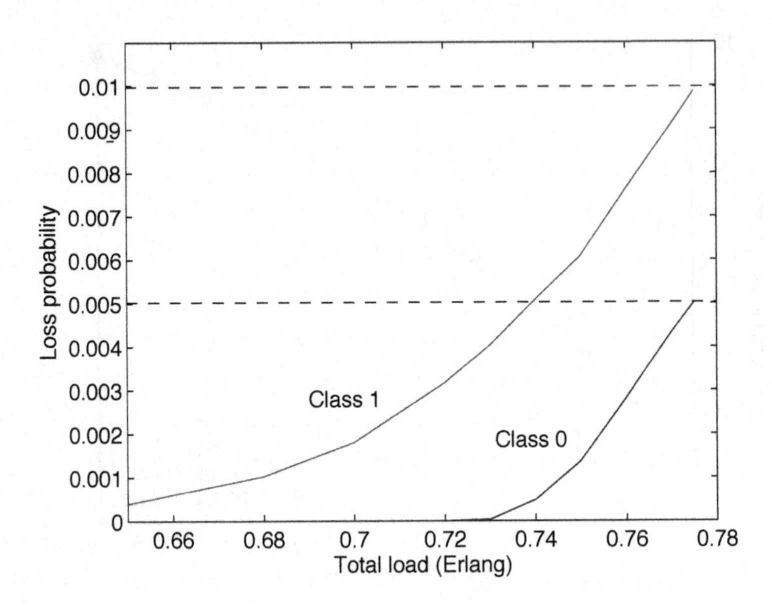

Fig. 4.4. Burst loss probabilities of individual classes vs total offered load

formance for individual flows within the same class as required by the second criterion in section 4.4.3.

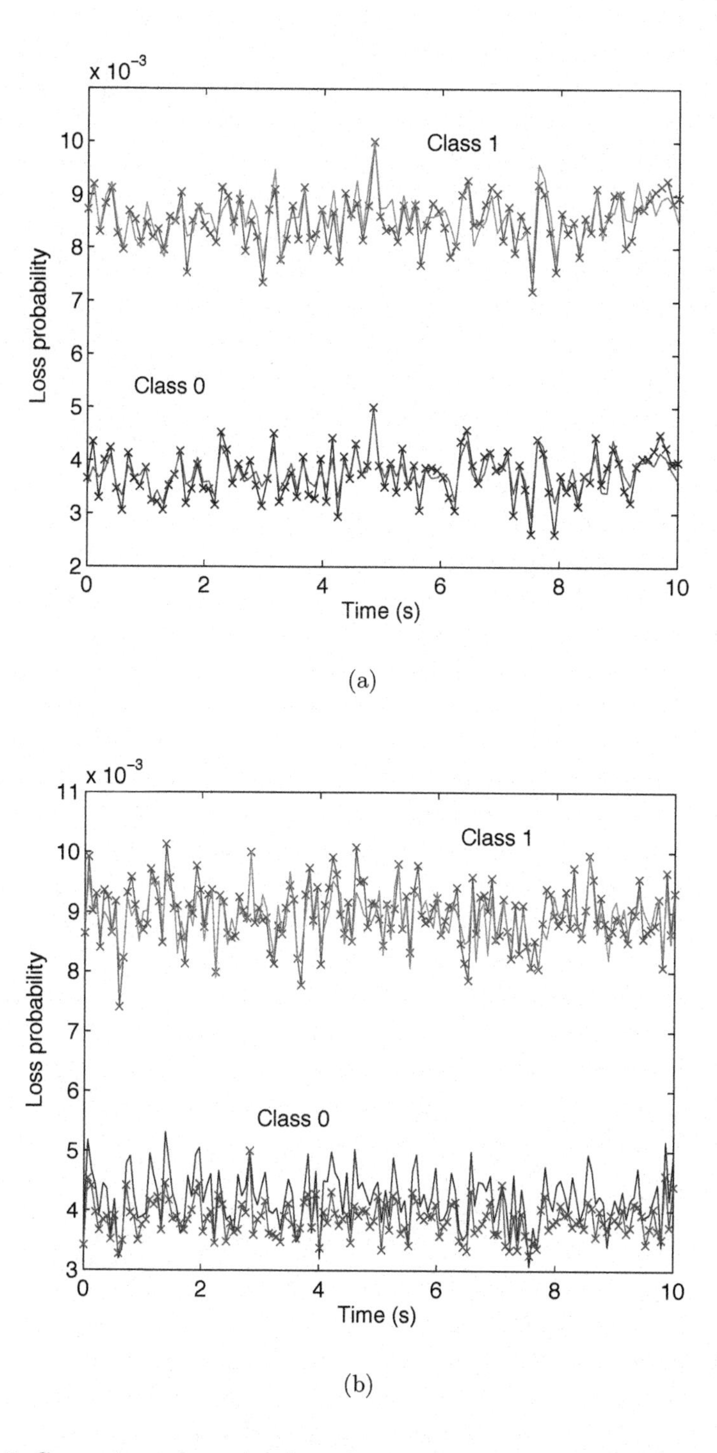

Fig. 4.5. Comparison of transient burst loss probabilities of traffic components with different characteristics: (a) Different offset times, and (b) Different burst lengths

References

1. Y. Chen, M. Hamdi, and D. Tsang, "Proportional QoS over OBS Networks," in *Proc. IEEE Globecom*, 2001, pp. 1510–1514.
2. Q. Zhang, V. M. Vokkarane, J. Jue, and B. Chen, "Absolute QoS Differentiation in Optical Burst-Switched Networks," *IEEE Journal on Selected Areas in Communications*, vol. 22, no. 9, pp. 1781–1795, 2004.
3. M. H. Phùng, K. C. Chua, G. Mohan, M. Motani, and T. C. Wong, "A Preemptive Differentiation Scheme for Absolute Loss Guarantees in OBS Networks," in *Proc. IASTED International Conference on Optical Communication Systems and Networks*, 2004, pp. 876–881.
4. M. H. Phùng, K. C. Chua, G. Mohan, M. Motani, and T. C. Wong, "An Absolute QoS Framework for Loss Guarantees in OBS Networks," *IEEE Transactions on Communications*, 2006, to appear.
5. H. Øverby and N. Stol, "Providing Absolute QoS in Asynchronous Bufferless Optical Packet/Burst Switched Networks with the Adaptive Preemptive Drop Policy," *Computer Communications*, vol. 28, no. 9, pp. 1038–1049, 2005.
6. X. Guan, I. L.-J. Thng, Y. Jiang, and H. Li, "Providing Absolute QoS through Virtual Channel Reservation in Optical Burst Switching Networks," *Computer Communications*, vol. 28, no. 9, pp. 967–986, 2005.

References

5

EDGE-TO-EDGE QOS MECHANISMS

The ultimate purpose of Quality of Service (QoS) mechanisms in a network is to provide end-to-end QoS to end users. To achieve this purpose, a wide range of mechanisms must be deployed in the network. They include both node-based mechanisms and core network edge-to-edge mechanisms. Node-based mechanisms such as burst scheduling and QoS differentiation have been discussed in previous chapters. In this chapter, edge-to-edge mechanisms within the core network to facilitate and realize end-to-end QoS provisioning, namely edge-to-edge QoS signalling and reservation mechanisms, traffic engineering mechanisms and fairness mechanisms will be discussed.

5.1 Edge-to-edge QoS Provisioning

Like node-based QoS differentiation mechanisms, edge-to-edge (e2e) QoS provisioning architectures can be categorised into *relative QoS* and *absolute QoS* models. In the relative QoS model, users are presented with a number of service classes such as Gold, Silver, Bronze. It is guaranteed that a higher priority class will always get a service that is no worse than that of a lower priority class. However, no guarantee is made on any quantitative performance metrics. If users have some applications that require quantitative guanrantees on some QoS metrics, different service classes will have to be tried out until the right one is reached. Apart from being inconvenient, such an approach cannot guarantee that the service level will be maintained throughout a session. This may be unacceptable to some users. Due to this reason, the relative QoS model, once popular in IP networks, is increasingly out of favour. To date, no comprehensive e2e QoS architecture based on the relative QoS model has been proposed for OBS networks.

On the other hand, absolute QoS architectures [1, 2, 3, 4] provide users with quantitative loss probabilities on an e2e basis. These architectures all divide the e2e loss budget, i.e., the maximum acceptable e2e loss probability, of a flow into small portions and allocate these to individual core nodes. The allocated loss thresholds are guaranteed by core nodes using some absolute QoS differentiation mechanisms. Node-based absolute QoS differentiation mechanisms have been discussed in the last chapter. This section will discuss and compare these architectures on an e2e basis.

5.1.1 Edge-to-edge classes as building blocks

An obvious way to realize e2e absolute QoS is to start with e2e QoS classes. In this approach [1, 2], network traffic is divided into e2e classes, each of which has an e2e loss budget. The loss budget for each class is then divided into equal portions, which are allocated to intermediate core nodes. Suppose that traffic class i has an e2e loss budget of $P_{C_i}^{NET}$ and the hop length that a flow f of the class has to traverse is H_f. The corresponding loss threshold allocated to each core node is given by

$$P_{C_i}^{MAX} = 1 - e^{ln(1-P_{C_i}^{NET})/H_f} = 1 - (1 - P_{C_i}^{NET})^{1/H_f}. \qquad (5.1)$$

A disadvantage of this approach is that one local threshold is required for every *(class, hop length)* pair. If the number of classes is N and the maximum hop length of all flows in the network is H, the number of local thresholds required will be $N_T = N \times H$. This is clearly not scalable for a large network where H is large.

To alleviate the above scalability problem, path clustering has been proposed in [2]. In this scheme, a number of path clusters are defined for the network. Each path cluster contains all paths of certain hop lengths. For example, for a network with a maximum hop length of 6, two clusters {1,2} and {3,4,5,6} may be defined. The first cluster contains all paths with hop lengths of 1 and 2. The remaining paths belong to the second cluster. Therefore, the number of local thresholds required is now $N_T = N \times H_c$ where H_c is the number of clusters. Since H_c can be much smaller than H, N_T is significantly reduced. The loss threshold for class i and cluster j is now given by

$$P_{C_i}^{MAX} = 1 - (1 - P_{C_i}^{NET})^{1/H_j^{MAX}} \tag{5.2}$$

where H_j^{MAX} is the maximum hop length of cluster j.

There are two parameters that define a path clustering: the number of clusters and the elements in each cluster. The number of clusters depends on the number of local thresholds a core node can support, which in turn depends on its processing capability. On the other hand, the assignment of paths into each cluster is up to the network administrator to decide. The way path clusters are defined can have a significant impact on the performance of the network. It is recommended that the optimal path clustering be found offline. In this process, all possible path clustering assignments are tried out. Two selection criteria are considered. First, the selected assignment must be able to satisfy the loss probability requirements of guaranteed traffic classes. Of all assignments that pass the first criterion, the one that gives the lowest loss probability for best effort traffic is the optimal clustering.

5.1.2 Per-hop classes as building blocks

Using e2e classes as building blocks, while simple, has two inherent drawbacks. The first one is that it is not efficient in utilizing network capacity. An operational network typically has some bottleneck links where a large number of paths converge. The larger the amount of traffic that these bottleneck links can support, the larger the amount of traffic that can be admitted to the network. However, the equal loss budget partitioning dictates that these bottleneck links provide the same local loss thesholds as other lightly loaded links. Therefore, less traffic can be supported than if more relaxed thresholds are allocated specifically to the bottleneck links. The second drawback is that the approach is not scalable. It requires a disproportionately large number of local loss thresholds to be supported by core nodes compared to the number of e2e classes offered by the network. Although path clustering helps to alleviate the problem to a certain extent, it does not solve it completely.

To solve the above problems, an entirely different approach that uses per-hop classes as building blocks has been proposed[1] [3, 4]. Its key idea is to define a limited number of per-hop absolute QoS classes[2] first and enforce their loss thresholds at each link. The network then divides the required e2e loss probability of the flow into a series of small loss probabilities and maps them to the available thresholds at the intermediate links on the path. When each intermediate node guarantees that the actual loss probability at its link is below the allocated loss probability, the overall e2e loss guarantee is fulfilled.

The QoS framework includes two mechanisms to enforce per-hop thresholds for individual flows, i.e., a preemptive absolute QoS differentiation mechanism and an admission control mechanism. The differentiation mechanism, which was discussed in the previous chapter, allows bursts from classes that are in danger of breaching their thresholds to preempt bursts from other classes. Thus, burst loss is shifted among the classes based on the differences between the thresholds and the measured loss probabilities of the classes. The differentiation mechanism is also designed such that individual flows within a single class experience uniform loss probability. Hence, even though it works at the class level, its threshold preserving effect extends to the flow level. The admission control mechanism limits the link's offered load to an acceptable level and thereby makes it feasible to keep the loss probabilities of all classes under their respective thresholds.

---

[1] Portions reprinted, with permission, from (M. H. Phùng, K. C. Chua, G. Mohan, M. Motani, and T. C. Wong, "Absolute QoS Signalling and Reservation in Optical Burst-Switched Networks," in *Proc. IEEE Globecom*, pp. 2009-2013) ©[2004] IEEE.

[2] In the rest of this chapter, the term "class" refers to per-hop QoS class unless otherwise specified.

For the mapping of classes over an e2e path, it is assumed that a label switching architecture such as Multi-Protocol Label Switching (MPLS) [5] is present in the OBS network. In this architecture, each header packet carries a label to identify the Label Switched Path (LSP) that it belongs to. When a header packet arrives at a core node, the node uses the header packet's label to look up the associated routing and QoS information from its Label Information Base (LIB). The old label is also swapped with a new one. Label information is downloaded to the node in advance by a Label Distribution Protocol (LDP). Such label switching architecture enables an LSP to be mapped to different QoS classes at different links.

An e2e signalling and reservation mechanism is responsible for probing the path of a new LSP and mapping it to a class at each intermediate link. When the LSP setup process begins, a reservation message that contains the requested bandwidth and the required e2e loss probability of the LSP is sent along the LSP's path toward the egress node. The message polls intermediate nodes on their available capacity and conveys the information to the egress node. Based on this information, the egress node decides whether the LSP's request can be accommodated. If the result is positive, an array of QoS classes whose elements correspond to the links along the path is allocated to the LSP. The class allocation is calculated such that the resulting e2e loss probability is not greater than that required by the LSP. It is then signalled to the intermediate core nodes by a returned acknowledgement message.

Existing LSPs are policed for conformance to their reservations at ingress nodes. When the traffic of an LSP exceeds its reserved traffic profile, its generated bursts are marked as out of profile. Such bursts receive best-effort service inside the network.

This approach to absolute QoS provisioning based on per-hop QoS classes effectively solves the problems of network utilization and scalability. By assigning the incoming LSP to different classes at different links based on the links' traffic conditions, it allows bottleneck links to support more traffic and thereby increases the amount of traffic that the entire network can support. Also, by combining the small number of predefined per-hop QoS classes, a much larger number of e2e service levels can be offered to users.

5.1.3 Link-based admission control

In the absolute QoS paradigm, traffic is guaranteed some upper bounds on the loss probabilities experienced at a link. Link-based admission control, which is used in the framework in section 5.1.2, is responsible for keeping the amount of offered traffic to a link in check so that the loss thresholds can be preserved. Since differentiation mechanisms shift burst loss among traffic classes at the link and keep the distances to thresholds of the classes equal, the admission control routine only needs to keep the average distance to threshold greater than zero. In other words, it needs to keep the overall loss probability smaller than the weighted average threshold. Suppose there are M QoS classes at the node and let T_i and B_i be the predefined threshold and the total reserved bandwidth of the ith class, respectively. The weighted average threshold is calculated as

$$T = \frac{\sum_{i=1}^{M} T_i B_i}{\sum_{j=1}^{M} B_j}. \tag{5.3}$$

The calculation of the overall loss probability P depends on the differentiation mechanism in use since different differentiation mechanisms will have different formulas to calculate the burst loss probability. In case that the formulas are not available, an empirical graph of the overall loss probability versus the total offered load may be used.

A reservation request will contain the amount of bandwidth to be reserved b_0 and the QoS class c to accommodate b_0 in. When a request arrives, the admission control routine substitutes the total reserved bandwidth B_c of class c with $B_c' = B_c + b_0$ and recalculates the weighted average threshold T' and the overall loss probability P' as above. If $P' \leq T'$, the request is admitted. Otherwise, it is rejected.

5.1.4 Per-hop QoS class definition

In the absolute QoS framework that uses per-hop QoS classes as building blocks to construct e2e loss guarantees, the definition of those classes is an important part of configuring the system. Usually, the number of classes M, which is directly related to the complexity of a core node's QoS differentiation block, is fixed. Hence, in this process, one only decides on where to place the available thresholds, namely the lowest and highest loss thresholds T_l and T_h and those between them.

Consider an OBS network in which LSPs have a maximum path length of H hops and a required e2e loss guarantee between P_l and P_h (not counting best-effort and out-of-profile traffic). The case requiring the lowest loss threshold T_l occurs when an LSP over the longest H-hop path requires P_l. Thus, T_l can be calculated as follows

$$T_l = 1 - (1 - P_l)^{1/H}. \tag{5.4}$$

Similarly, the highest threshold is $T_h = P_h$ for the case when a one-hop LSP requires P_h.

When considering how to place the remaining thresholds between T_l and T_h, it is noted that since the potential required e2e loss probability P_0 is continuous and the threshold values are discrete, the e2e loss bound P_{e2e} offered by the network will almost always be more stringent than P_0. This "discretization error" reduces the maximum amount of traffic that can be admitted. Therefore, the thresholds need to be spaced so that this discretization error is minimised. A simple and effective way to do this is to distribute the thresholds evenly on the logarithmic scale. That is, they are assigned the values T_l, γT_l, $\gamma^2 T_l, \ldots, \gamma^{M-1} T_l$, where $\gamma = (T_h/T_l)^{1/(M-1)}$.

5.1.5 Edge-to-edge signalling and reservation

Edge-to-edge signalling and reservation mechanisms, as the name implies, are responsible for coordinating the QoS reservation setup and teardown for LSPs over the e2e paths. Among the e2e absolute QoS proposals in the literature, only the framework discussed in 5.1.2 includes a comprehensive e2e signalling and reservation mechanism. It is described and discussed below.

During the reservation process of an LSP, the signalling mechanism polls all the intermediate core nodes about the remaining capacity on the output links and conveys the information to the egress node. Using this information as the input, the egress node produces a class allocation that maps the LSP to an appropriate class for each link on the path. The signalling mechanism then distributes the class allocation to the core nodes. As a simple illustration, consider an LSP with an e2e loss requirement of 5% that needs to be established over a 4-hop path and the second hop is near congestion. Suppose the lowest threshold is $T_l = 0.05\%$ and the ratio between two adjacent thresholds is $\gamma = 2$. The network allocates the LSP a threshold of $\gamma^6 T_l = 3.2\%$ for the second hop and $\gamma^3 T_l = 0.4\%$ for the other hops to reflect the fact that the second node is congested. The resulting guaranteed upper bound on e2e threshold will roughly be 4.4%, satisfying the LSP's requirement.

The QoS requirements of an LSP consists of its minimum required bandwidth and its maximum e2e loss probability. As new IP flows join an LSP or existing IP flows terminate, a reservation or teardown process needs to be carried out for the LSP. The reservation scenarios for an LSP can be categorised as follows.

1. A new LSP is to be established with a specified minimum bandwidth requirement and a maximum e2e loss probability. This happens when some IP flow requests arrive at the ingress node and cannot be fitted into any of the existing LSPs.
2. An existing LSP needs to increase its reserved bandwidth by a specified amount. This happens when some incoming IP flows have e2e loss requirements compatible with that of the LSP.
3. An existing LSP needs to decrease its reserved bandwidth by a specified amount. This happens when some existing IP flows within the LSP terminate.
4. An existing LSP terminates because all of its existing IP flows terminate.

The detailed reservation process for the first scenario is as follows. The ingress node sends a reservation message towards the egress node over the path that the LSP will take. The message contains a requested bandwidth b_0 and a required e2e loss probability P_0. When a core node receives the message, its admission control routine checks each class to see if the requested bandwidth can be accommodated in that class. The check starts from the lowest index class, which corresponds to the lowest threshold, and moves up. The node stops at the first satisfactory class and records in the message the class index c and a parameter κ calculated as follows

$$\kappa = \left\lfloor log_{\gamma}\left(\frac{T'}{P'}\right)\right\rfloor \tag{5.5}$$

where T' and P' are as described in section 5.1.3 and γ is the ratio between the thresholds of two adjacent classes. These parameters will be used by the egress node for the final admission control and class allocation. The message is then passed downstream. The node also locks in the requested bandwidth by setting the total reserved bandwidth B_c of class c as $B_c(new) = B_c(old) + b_0$ so that the LSP will not be affected by later reservation messages. On the other hand, if all the classes have been checked unsuccessfully, the request is rejected and an error message is sent back to the ingress node. Upon receiving the error message, the upstream nodes release the bandwidth locked up earlier.

The final admission control decision for the LSP is made at the egress node. The received reservation message contains two arrays $\boldsymbol{c}$ and $\boldsymbol{\kappa}$ for the intermediate links of the path. Assuming burst loss at each link is independent, the lowest possible e2e loss probability P^0_{e2e} given as

$$P^0_{e2e} = 1 - \prod_{i=1}^{n}(1 - p^0_i) \tag{5.6}$$

where p^0_i is the lowest threshold offered by the ith node on a n-node path. If $P^0_{e2e} \leq P_0$, the request is admitted. The egress node then allocates each core node one of the predefined classes to support the LSP such that

$$\begin{cases} p_i \geq p_i^0 \\ P_{e2e} = 1 - \prod_{i=1}^{n}(1 - p_i) \leq P_0 \end{cases} \tag{5.7}$$

where p_i is the threshold of the class allocated to the LSP at the ith node and P_{e2e} is the corresponding e2e loss probability. The class allocation algorithm will be described in the next section. This class allocation is signalled back to the intermediate core nodes using a returned acknowledgement message that contains the old index array $\boldsymbol{c}$ and an allocated index array $\boldsymbol{c_a}$. Upon receiving the acknowledgement message, a core node moves the reserved bandwidth of the LSP from class c to class c_a. The new LSP is allowed to start only after the ingress node has received the successful acknowledgement message. If $P_{e2e}^0 > P_0$, the request is rejected and an error message is sent back to the ingress node. The intermediate core nodes will release the locked bandwidth upon receiving the error message.

The reservation process for the second scenario is relatively simpler. In this case, the ingress node sends out a reservation message containing the requested bandwidth b_0 and the LSP's label. Since there is already a QoS class associated with the LSP at each of the core nodes, a core node on the path only needs to check if b_0 can be supported in the registered class. If the outcome is positive, the node locks in b_0 and passes the reservation message on. Otherwise, an error message is sent back and the upstream nodes release the bandwidth locked previously. If the reservation message reaches the egress node, a successful acknowledgement message is returned to the ingress node and the LSP is allowed to increase its operating bandwidth.

In the last two scenarios, the reservation processes are similar. The ingress node sends out a message carrying the amount of bandwidth with a flag to indicate that it is to be released and the LSP's label. The released bandwidth is equal to the reserved bandwidth of the LSP if the LSP terminates. At intermediate core nodes, the total reserved bandwidth of the class associated with the LSP is decreased by that amount. No admission control check is necessary. Since the core nodes do not keep track of bandwidth reservation by individual LSPs, the processing at core nodes is identical for both scenarios. It should be noted that when an LSP terminates, there is a separate signalling process to remove the LSP's information from core nodes' LIBs. However, it is not considered part of the QoS framework.

5.1.6 Dynamic class allocation

In the above signalling and reservation mechanism, when an egress node has determined that an LSP request is admissible, it uses a dynamic class allocation algorithm to find the bottleneck link and allocate the class with the highest possible threshold to it while still keeping the existing traffic below their respective thresholds. This class allocation shifts some of the loss guarantee burden from the bottleneck link to other lightly loaded links. Since the remaining capacity of the path is determined by the bottleneck link, the algorithm will maximise the path's remaining capacity and allow more future QoS traffic to be admitted.

For this purpose, the egress node has at its disposal two arrays $\boldsymbol{c}$ and $\boldsymbol{\kappa}$ recorded in the reservation message by the core nodes. As described in the last section, $\boldsymbol{c}[i]$ is the lowest index class (with the lowest threshold) that can accommodate the requested bandwidth at the ith node. As long as $\boldsymbol{c}[i] > 0$, it is an accurate indicator of how much capacity is available at link i since it cannot be decreased further without making P' exceed T'. However, when $\boldsymbol{c}[i] = 0$, how much lower P' is compared to T' is not known based on $\boldsymbol{c}[i]$ alone. Hence, $\boldsymbol{\kappa}[i]$ given by (5.5) is introduced to enable the core node to convey that information to the egress node. It is observed that when $\boldsymbol{c}[i] > 0$, $\gamma^{-1}T' < P' \leq T'$. Therefore, $\boldsymbol{\kappa}[i] > 0$ only if $\boldsymbol{c}[i] = 0$. Thus, $\boldsymbol{c}-\boldsymbol{\kappa}$ indicates the remaining capacity at the intermediate links in all cases. The higher the value of $\boldsymbol{c}[i] - \boldsymbol{\kappa}[i]$, the lower the remaining capacity at link i and vice versa.

Based on the above observation, the class allocation algorithm is detailed in Algorithm 1. In executing this algorithm, negative class indices in $\boldsymbol{c_a}$ are counted as zero. In the first two lines, the algorithm sets $\boldsymbol{c_a}$ such that the maximum element is $M - 1$ and the differences among the elements are the same as in the array

$\boldsymbol{c} - \boldsymbol{\kappa}$. Next, it repeatedly decrements all the elements of $\boldsymbol{c_a}$ until $P_{e2e} \leq P_0$. Finally, the elements of $\boldsymbol{c_a}$ are incremented one by one until just before $P_{e2e} > P_0$ in order to push P_{e2e} as close as possible to P_0 without exceeding it.

Algorithm 1: Class allocation algorithm

(1) $d_{max} \leftarrow \text{MAX}(\boldsymbol{c} - \boldsymbol{\kappa})$
(2) $\boldsymbol{c_a} \leftarrow (\boldsymbol{c} - \boldsymbol{\kappa}) - d_{max} + M - 1$
(3) **while** $P_{e2e} > P_0$
(4) $\quad \boldsymbol{c_a} \leftarrow \boldsymbol{c_a} - 1$
(5) Increment elements of $\boldsymbol{c_a}$ one by one until just before $P_{e2e} > P_0$

As an illustrative example, consider an OBS network that has 8 predefined QoS classes associated with each link. The class indices are {0, 1, ..., 7}. The lowest threshold is $T_l = 0.05\%$ and the ratio between two adjacent thresholds is $\gamma = 2$. These settings provide a sufficiently small lowest threshold T_l and sufficiently fine threshold increment to satisfy most typical e2e loss requirements. An LSP with an e2e loss requirement of 1% is to be set up over a three-hop path. Its required bandwidth is assumed to be very small compared to the link capacity. The utilisation levels at the intermediate links are {0.3, 0.6, 0.35}. Suppose the received message at the egress node contains $\boldsymbol{c} = \{0, 0, 0\}$ and $\boldsymbol{\kappa} = \{50, 2, 35\}$. The values of $\boldsymbol{\kappa}$ indicate that although all links are relatively lightly loaded, the second link has the least available capacity. Therefore, the class allocation algorithm should give it the highest threshold. Going through the algorithm, $\boldsymbol{c_a} = \{-41, 7, -26\}$ on line (3) and $\boldsymbol{c_a} = \{-44, 7, -29\}$ on line (5). The final result is $\boldsymbol{c_a} = \{1, 4, 1\}$ corresponding to thresholds of {0.1%, 0.8%, 0.1%}. This shows that the algorithm successfully allocates the maximum possible class index to the bottleneck node.

Most of the computations in the above algorithm are array manipulation and the calculation of P_{e2e} is according to (5.6). Since their computational complexity is determined by the number of intermediate core nodes, which is usually small, the computational complexity of the algorithm is not significant.

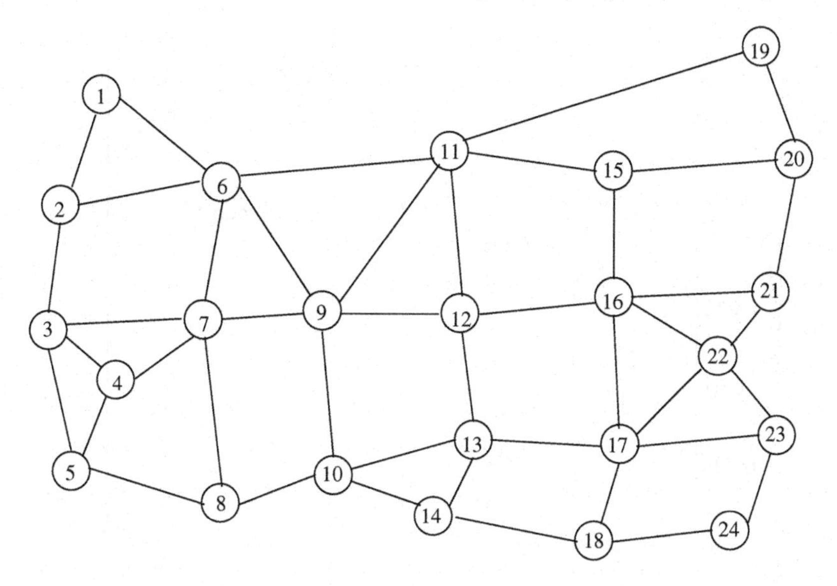

Fig. 5.1. 24-node NSF network topology

5.1.7 Numerical study

In [3, 4], simulation is used to evaluate the performance of the e2e absolute QoS framework described in 5.1.2. The topology in Figure 5.1, which is a simplified topology for the NSF backbone network, is used. It consists of 24 nodes and 43 links. Fixed shortest path routing is used and the maximum path length is 6. For simplicity, the propagation delays between adjacent nodes are assumed to have a fixed value of 5 ms. The links are bi-directional, each implemented by two uni-directional links in opposite directions. Each uni-directional link has an output link with 64 data wavelengths, each having a transmission rate of 10 Gbps. Each node is assumed to have full wavelength conversion capability and no buffer. Bursts arrive at a link according to a Poisson process with rate λ. Seven per-hop QoS classes are defined at each link with the lowest threshold $T_l = 0.0005$ and the ratio between two adjacent thresholds $\gamma = 2$.

New LSPs are generated at each node according to a Poisson process with rate λ_{LSP} and have exponentially distributed durations. For simplicity, it is assumed that LSPs do not change their bandwidth requirements. Two groups of LSPs are consid-

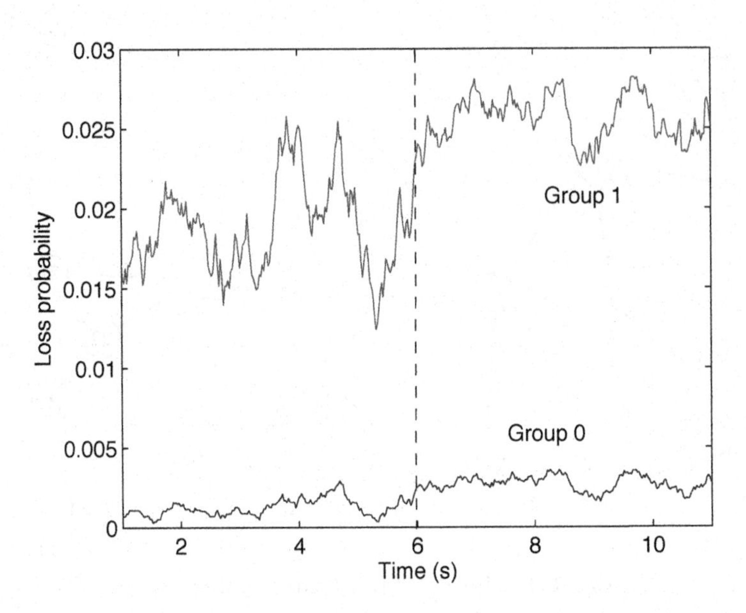

Fig. 5.2. Transient edge-to-edge burst loss probability of two traffic groups with e2e loss requirements of 0.01 and 0.05

ered: group 0 and group 1 with required e2e loss probabilities of 0.01 and 0.05, respectively. A new LSP is destined to a random node and falls into one of the two groups with equal probability.

In one experiment, the temporal loss behaviour of the framework is examined. To do this, the simulation is run for 11 s and the e2e loss rate of traffic between node pair (1, 24) is monitored. The path between this node pair is 6 hops long, which is the longest path in the network, and it runs through the bottleneck link (9,10). The data in the first second, which is the system warm-up period, is discarded. During the first 6 s, the total network load is 15 Erlang, which is equally distributed among all node pairs. After that, the offered load between node pair (1,24) is increased 10 folds. The loss rates of the two groups are plotted against time in Figure 5.2. It is observed that the loss probabilities increase in response to the increase in the offered load at $t = 6$ s. Nevertheless, they are always kept below the respective thresholds. This shows that the reservation process is able to guarantee the loss bounds to admitted LSPs in real time regardless of the traffic load.

Another observation from Figure 5.2 is that the maximum loss probabilities of the two traffic groups are 0.004 and 0.03, which are well below the required e2e loss probabilities. This is due to the fact that almost all of the burst loss on the path occurs at a single bottleneck link. Hence, the e2e loss probabilities are limited by the maximum thresholds that can be allocated to the bottleneck link. In this case, they are 0.004 and 0.032, respectively. If more per-hop classes are available, the gaps between adjacent thresholds will be reduced and the e2e loss probabilities can be pushed closer to the targets.

In Figure 5.3, the e2e loss probabilities of LSPs with different hop lengths and at two different loads of 15 and 30 Erlang are plotted. The same loss probabilities of the path clustering scheme proposed in [2] are also plotted for comparison. No admission control implementation is provided in [2] for the path clustering scheme. The cluster combination {1,2}{3,4,5,6} is used as it is the best performing one. It groups LSPs with one or two hop lengths into one cluster and all the remaining LSPs into the other cluster.

A number of observations can be made from Figure 5.3. Firstly, the e2e loss probabilities of all LSP groups are below their required e2e levels. This is true under both medium and heavy loading conditions. Secondly, the loss probabilities increase from 1-hop group to 3-hop group but level off after that. The loss probability increase is due to the fact that burst traversing more hops will experience more loss. However, at a certain level, the effect of admission control dominates and the loss probabilities stay nearly constant. For the path clustering scheme, it is observed that it can keep the e2e loss probabilities for group 0 LSPs below the required level. However, this is achieved at great cost to group 1 LSPs, which experience very high loss probabilities. This happens because there is no admission control present, so core nodes must drop low priority bursts excessively in order to keep within the loss guarantees for high priority bursts. Another observation is that the loss probabilities of group 0 LSPs in the path clustering scheme vary significantly with hop lengths. This is because the scheme allocates the same per-hop threshold to LSPs within a cluster. Therefore, LSPs in a cluster with many different hop lengths such as {3,4,5,6}

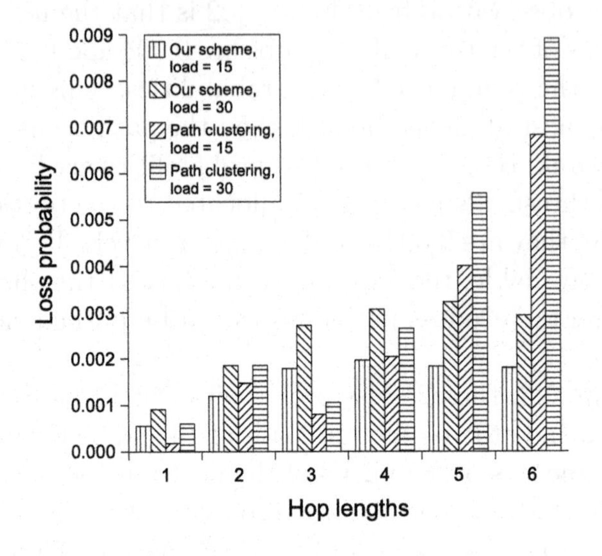

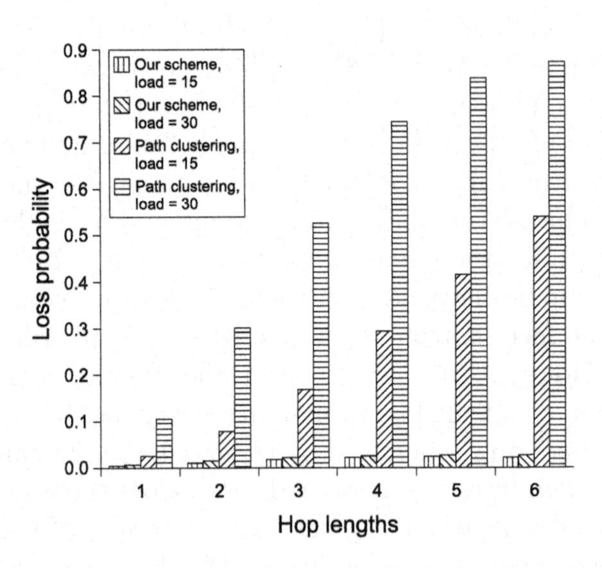

Fig. 5.3. Average e2e loss probability of LSPs with different hop lengths for our scheme and path clustering scheme: (a) Traffic group 0 (required e2e loss probability of 0.01), and (b) Traffic group 1 (required e2e loss probability of 0.05)

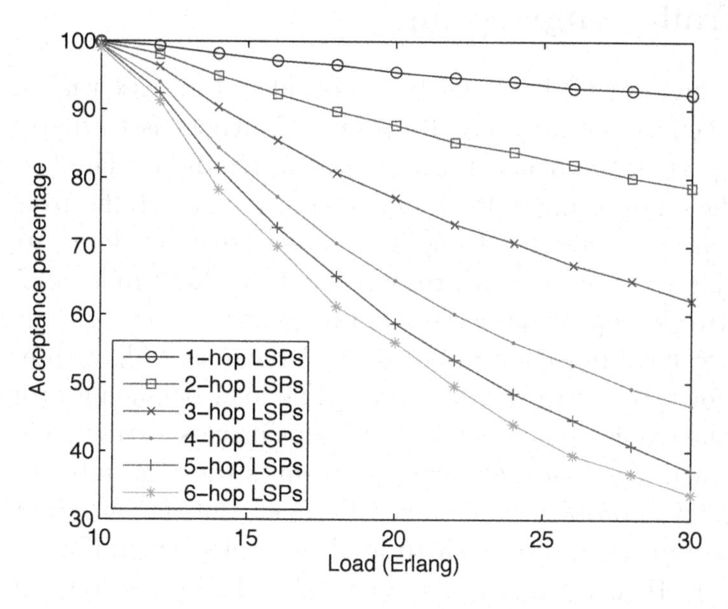

Fig. 5.4. Overall acceptance percentage of LSPs with different hop lengths versus average network load

will experience significantly different e2e loss probabilities, some of which are far below the required level.

In the final experiment, the acceptance percentage of LSP groups with different hop lengths are plotted against the network loads in Figure 5.4. It shows that the acceptance percentage of all LSP groups decrease with increasing load, which is expected. Among the groups, the longer the hop length, the worse the performance. There are two reasons for this. Firstly, the network must allocate more stringent per-hop thresholds to longer LSPs compared to shorter LSPs that have the same required e2e loss probability. Secondly, longer LSPs are more likely to encounter congestion on one of their intermediate links. This situation can be remedied by a fairness scheme that gives more favourable treatment to longer LSPs in the reservation process.

5.2 Traffic Engineering

Traffic engineering has long been considered an essential part of a next-generation network. Its primary purpose is to map traffic to the part of the network that can best handle it. The key form of traffic engineering is load balancing, in which traffic from congested areas is diverted to lightly loaded areas. In doing so, load balancing frees up network resource at bottleneck links and helps the network to provide better QoS to users.

There have been a number of load balancing algorithms proposed for OBS networks. In early offline load balancing proposals [6, 7], the traffic demands among various ingress/egress node pairs are known. The load balancing problem is then formulated as an optimization problem and solved by linear integer programming. This optimization approach to load balancing is similar to what is done in IP networks. Later proposals take into account unique features of OBS networks to improve performance. In this section, these algorithms will be presented.

5.2.1 Load balancing for best effort traffic

In [8, 9], a load balancing scheme[3] for best effort traffic in OBS networks is proposed independently by two research groups. The load balancing scheme is based on *adaptive alternative routing.* For each node pair, two link-disjoint alternative paths are used for data transmission. Label switched paths (LSPs) for the above pre-determined paths could be set up to facilitate transmission of header packets with reduced signaling and processing overhead. For a given node pair, traffic loads which are the aggregation of IP flows arrive at the ingress node and are adaptively assigned to the two paths so that the loads on the paths are balanced. A time-window-based mechanism is adopted in which adaptive alternate routing operates in cycles of specific time duration called time windows. Traffic assignment on the two paths are periodically

---

[3] Reprinted from (J. Li, G. Mohan, and K. C. Chua, "Dynamic Load Balancing in IP-over-WDM Optical Burst Switching Networks," *Computer Networks*, vol. 47, no. 3, pp. 393–408), ©[2005], with permission from Elsevier.

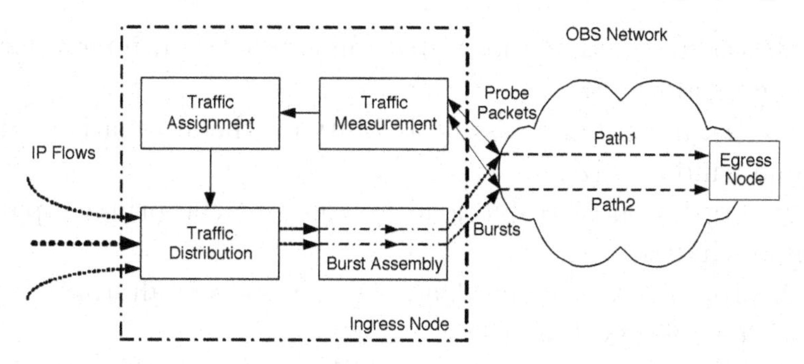

Fig. 5.5. Functional blocks of the best-effort load balancing scheme

adjusted in each time window based on the statistics of the traffic measured in the previous time window.

Figure 5.5 shows the functional block diagram of the load balancing scheme for a specific node pair. At the ingress node, four functional units – traffic measurement, traffic assignment, traffic distribution and burst assembly – work together to achieve load balancing. Traffic measurement is responsible for collecting traffic statistics by sending probe packets to each of the two paths periodically. The collected information is then used to evaluate the impact of traffic load on the two paths. Based on the measurements and the hop difference between the two alternative paths, traffic assignment determines the proportion of traffic allocated to each of the two paths in order to balance the traffic loads on the two paths by shifting a certain amount of traffic from the heavily-loaded path to the lightly-loaded path. Traffic distribution plays the role of distributing the IP traffic that arrives at the ingress node to the two paths according to the decisions made by traffic assignment. Finally, bursts are assembled from packets of those flows assigned to the same path. The processes are described in detail below.

For ease of exposition, the following are defined:

- $path_p$: primary path.
- $path_a$: alternative path.
- $length_p$: hop count of the primary path.
- $length_a$: hop count of the alternative path.
- $T(i)$: ith time window.

- $loss_p(i)$: mean burst loss probability on the primary path in time window $T(i)$.
- $loss_a(i)$: mean burst loss probability on the alternative path in time window $T(i)$.
- P_p^i: proportion of traffic load assigned to the primary path in time window $T(i)$.
- P_a^i: proportion of traffic load assigned to the alternative path in time window $T(i)$.
- $(P_p, P_a)^i$: combination of P_p^i and P_a^i which represents the traffic assignment in time window $T(i)$.

Note that $length_p \leq length_a$ and $P_p^i + P_a^i = 1$.

Traffic measurement

The traffic measurement process is invoked periodically in each time window. The 'mean burst loss probability' is used as the measured performance metric. The purpose of traffic measurement is to collect traffic statistics for each path by sending probe packets and then calculating the mean burst loss probability to evaluate the impact of traffic load. Since the traffic measurement process is similar in each time window, the following describes the whole process for a specific time window $T(i)$ only.

At the beginning of $T(i)$, the ingress node starts recording the total number of bursts $total(s, path_p)$ and $total(s, path_a)$ sent to each path $path_p$ and $path_a$, respectively. At the end of $T(i)$, it sends out probe packets to each path to measure the total number of dropped bursts $dropped_p$ and $dropped_a$ on each path during $T(i)$. After receiving the successfully returned probe packets, it updates $loss_p(i)$ and $loss_a(i)$ as follows:

$$loss_p(i) = \frac{dropped_p}{total(s, path_p)};$$

$$loss_a(i) = \frac{dropped_a}{total(s, path_a)}.$$

At each intermediate node, a burst loss counter is maintained. At the beginning of $T(i)$, the counter is reset to zero. It is incremented every time a burst is lost at the node. When the probe packet arrives, the node adds the current value of the counter to the burst loss sum carried by the probe packet and records the new value to the probe packet.

Finally, after receiving the probe packets, the egress node returns them to the ingress nodes as successful probe packets.

Traffic assignment

Traffic assignment adaptively determines the proportion of traffic allocated to each of the two paths in each time window. The traffic assignment decision is determined by two parameters: the measured values of the mean burst loss probability on the two paths and the hop count difference between the two paths. The measured mean burst loss probabilities returned by traffic measurement in the previous time window are used to estimate the impact of traffic loads on the two paths. These loads are balanced in the current time window. The basic idea is to shift a certain amount of traffic from the heavily-loaded path to the lightly-loaded path so that traffic loads on the two paths are balanced.

Hop count is an important factor in OBS networks for the following two reasons:

1. Since burst scheduling is required at each intermediate node traversed, a longer path means a higher possibility that a burst encounters contention.
2. A longer path consumes more network resources which results in a lower network efficiency.

Thus, network performance may become poorer if excessive traffic is shifted from the shorter path to the longer path even though the longer path may be lightly loaded. To avoid this, a protection area PA is set whose use is to determine when traffic should be shifted from the shorter path ($path_p$) to the longer path ($path_a$). Let the measured mean burst loss probability difference between the two paths ($loss_p(i) - loss_a(i)$) be Δp. If and only if Δp is beyond PA, traffic can be shifted from the shorter path ($path_p$) to the longer path ($path_a$). Let the hop count difference between the two paths ($length_a - length_p$) be Δh. PA is given by $PA = \Delta h \times \tau$, where τ is a system control parameter. Thus, a good tradeoff is achieved between the benefit of using a lightly-loaded path and the disadvantage of using a longer path.

Consider the traffic assignment process in a specific time window $T(i)$. Initially, in time window $T(0)$, the traffic is distributed in the following way:

$$P_p^0 = \frac{length_a}{length_p + length_a}, \tag{5.8}$$

$$P_a^0 = \frac{length_p}{length_p + length_a}. \tag{5.9}$$

Let the mean burst loss probabilities of the two paths returned by traffic measurement in time window $T(i-1)$ be $loss_p(i-1)$ and $loss_a(i-1)$, respectively. Then $\Delta p = loss_p(i-1) - loss_a(i-1)$. Let the traffic assignment in time window $T(i-1)$ be $(P_p, P_a)^{i-1}$. The following procedure is used to determine $shiftP$ (the amount of traffic to be shifted) and the new traffic assignment $(P_p, P_a)^i$ in time window $T(i)$.

1. If $\Delta p \geq PA$, then traffic is shifted from $path_p$ to $path_a$,
   $shiftP = P_p^{i-1} \times (\Delta p - PA)$;
   $P_p^i = P_p^{i-1} - shiftP$;
   $P_a^i = P_a^{i-1} + shiftP$;
   else if $\Delta p < PA$ and $\Delta p \geq 0$, then traffic assignment remains the same,
   else if $\Delta p < 0$, then traffic is shifted from $path_a$ to $path_p$,
   $shiftP = P_a^{i-1} \times \mid \Delta p \mid$;
   $P_p^i = P_p^{i-1} + shiftP$;
   $P_a^i = P_a^{i-1} - shiftP$;
   end if.
2. Send the new traffic assignment information to the traffic distribution unit.
3. At the end of time window $T(i)$, receive the values of $loss_p(i)$ and $loss_a(i)$ from the traffic measurement unit.
4. Let i=i+1 and go to step 1.

Traffic distribution

The traffic distribution function distributes IP flows arriving at the ingress node to the two paths based on the traffic assignment decision. One way to distribute the traffic is on a per-burst basis. Here, bursts are mapped to one of the paths to achieve load balancing. This means that packets from a given flow could be in bursts traversing different paths of different lengths. This may require the reordering of packets at the egress node which is undesirable. Another possible way is to distribute traffic on a per-flow basis. Once a flow is distributed to a path, the packets belonging to the flow are transmitted on this path. This means that the packets from those flows which are mapped to the same path are assembled into bursts. Unlike the per-burst based traffic distribution method, the order of packets is mostly preserved in the flow-based distribution method. Reordering of packets is needed only if the flows are shifted from a longer-delay path to a shorter-delay path when traffic assignment is adjusted. Therefore, flow-based traffic distribution is more desirable and is used in this scheme.

Path selection

The purpose of alternative-path selection is to choose two link-disjoint paths, primary path $path_p$ and alternative path $path_a$, for each node pair to be used by the adaptive alternate routing algorithm. The alternative paths for each node pair are pre-determined and the LSPs are set up accordingly. Two schemes are considered: *shortest-hop path routing* (SHPR) and *widest-shortest-hop path routing* (WSHPR). The key idea of the path selection scheme is to associate cost metrics to links in a certain way and use a Dijkstra-like minimum-cost path selection algorithm to optimize a certain path-cost metric.

In SHPR, only hops are considered as the cost metric and the path with the minimum number of physical links is chosen. By assigning a weight of 1 to all the links and running a shortest-path selection algorithm such as Dijkstra's algorithm, the primary path $path_p$ is chosen first. Then, every link l_i constituting the primary path is removed from the set of all links E to get a residual link set E'. By running the shortest-path routing algorithm on E', the alternative path $path_a$ is chosen such that each constituted link of the path belongs to E'.

In WSHPR, both hops and link load are considered as the cost metrics. WSHPR chooses a path that minimizes the maximum link load among all the shortest-hop paths. The cost of a link l_i has two components, k_1 and $k_2Tl(i)$, where k_1 and k_2 are constants and $Tl(i)$ is the long term average load on link l_i. The positive constants k_1 and k_2 are chosen such that $k_1 >> k_2$. Dijkstra's algorithm can suitably be modified to choose the minimum cost path where the cost of a path is defined as

$$path\_cost = \sum_{i=1}^{I} k_1 + max_{i=1}^{I} k_2 \times Tl(i) \tag{5.10}$$

where I is the length of the path. The primary and alternative paths are then selected in a similar manner to that in SHPR.

Performance study

In [9], the adaptive alternative routing algorithm (AARA) is studied via simulation. The simulation uses the Pan-European backbone network topology as shown in Figure 5.6. Each link consists of two uni-directional fibres. Each fibre contains 4 wavelengths, each with capacity of 1 Gbps. A basic void-filling scheduling algorithm [10] is used to schedule bursts. The burst assembler generates a burst every 80 μs, which is chosen to yield an average burst duration of about 10 μs with moderate traffic load on the 1 Gbps data channel.

A *Long Range Dependent* (LRD) traffic model is used in the study. In this LRD traffic model, traffic that arrives at each node pair in the network is the aggregation of multiple IP flows [11]. Each IP flow is an ON/OFF process with Pareto distributed ON and OFF times. During each ON period of the Pareto-ON/OFF model, a Pareto distributed number of packets, with mean N and Pareto shape parameter β, are generated at some peak rate p packets/sec. The OFF times are also Pareto distributed with mean I and shape parameter γ. The following values are used for the Pareto-ON/OFF flows in the simulations: $N = 5$, $\beta = 1.2$, $I =$

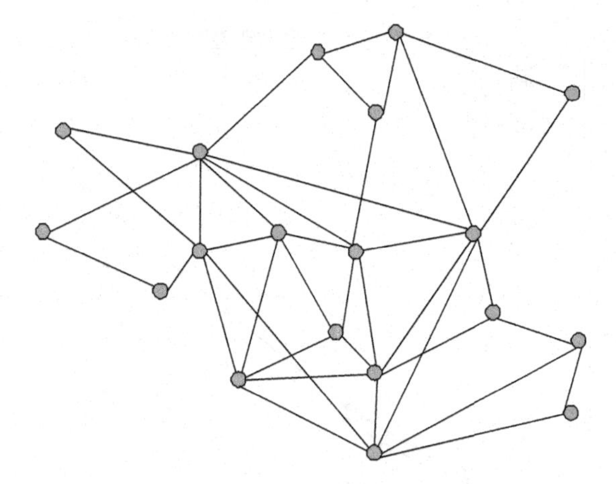

Fig. 5.6. Pan-European optical backbone network

$56,000\ \mu s$, $\gamma = 1.1$, $p = 640$. The packet length is assumed to be 400 *bytes*, which is the average packet length used in [10]. Given the above parameters, the mean transmission rate per flow r can be approximately calculated as 80 kb/s. IP flows are assumed to arrive at a node pair according to a Poisson process. The system control parameter τ is set to 0.01. It is used to avoid excessive traffic shifting from the shorter path to the longer path.

Two other algorithms, *shortest-hop path routing - fixed routing* (SHPR-FR) and *static alternate routing* (SAR) are also implemented as benchmarks. In SHPR-FR all the bursts are transmitted along the fixed shortest path. In SAR bursts are transmitted along the two pre-determined alternative paths which are chosen as the first shortest-hop path and the link-disjoint next shortest-hop path. It differs from AARA in that the traffic flows are distributed between the two paths with the same initial traffic assignment as AARA but remain unchanged. Since the choice of alternative path selection scheme has an impact on the performance of the load balancing algorithms, different alternative path selection schemes are applied to AARA and SAR. That is, two AARA algorithms based on the two different alternative path selection schemes, *SHPR-based AARA* (SHPR-AARA) and *WSHPR-based AARA* (WSHPR-AARA) and two SAR algorithms, *SHPR-based*

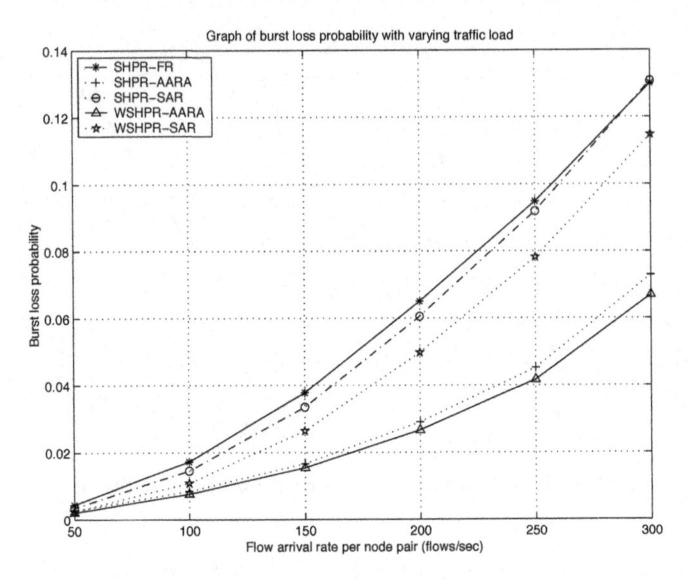

Fig. 5.7. Burst loss probability versus traffic load

SAR (SHPR-SAR) and *WSHPR-based SAR* (WSHPR-SAR) are studied in the simulations.

Figure 5.7 shows the burst loss probability with varying traffic load per node pair for time window size $T = 500,000$ μs. The AARA algorithms, SHPR-AARA and WSHPR-AARA, perform much better than the SHPR-FR and SAR algorithms (SHPR-SAR and WSHPR-SAR). The performance is improved by up to 59% in comparison with SHPR-FR. Since AARA performs load balancing in the network, congestion is reduced. As a result, the number of bursts dropped due to contention is reduced. On the other hand, although algorithm SHPR-FR always chooses the shortest path to use, the number of dropped bursts is larger than that of AARA which may choose longer paths. Although SAR distributes traffic to the two paths, it performs worse than AARA because it fails to keep track of the different congestion states on the link-disjoint paths.

In general, the algorithms based on WSHPR perform better than the algorithms based on SHPR. This is because WSHPR considers the bottleneck link in the network and avoids congesting such links while determining the alternative paths. However,

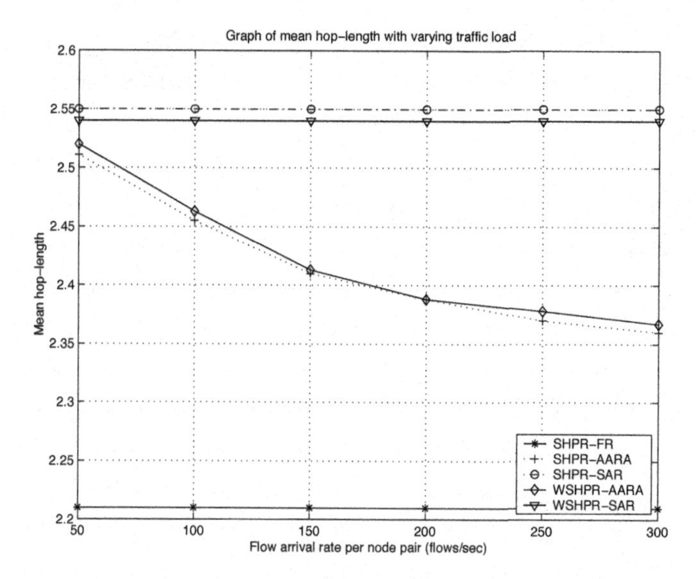

Fig. 5.8. Mean hop length versus traffic load

WSHPR requires prior knowledge of the traffic demands of the whole network which may not adapt to the dynamic traffic demands in the real network. WSHPR-AARA performs the best among all the algorithms since it considers load balancing through path selection as well as dynamic load balancing through adaptive load distribution. However, WSHPR-AARA may not perform well when the traffic demands deviate significantly from the estimated long-term average demands. On the other hand, SHPR-AARA preserves the advantage of dynamic adaptation without requiring a priori traffic load information and at the same time its performance is fairly good.

Figure 5.8 shows the mean hop-length traversed by a burst with varying traffic load per node pair. The mean hop-length affects the delay, signaling overhead and initial offset time in the network. The mean hop-length for the AARA algorithms, SHPR-AARA and WSHPR-AARA, is slightly larger than that for SHPR-FR . This implies that the additional delay, signaling overhead and initial offset time introduced by AARA is rather low when compared to the performance improvement achieved. The mean hop-length for the AARA algorithms also decreases when traffic load increases.

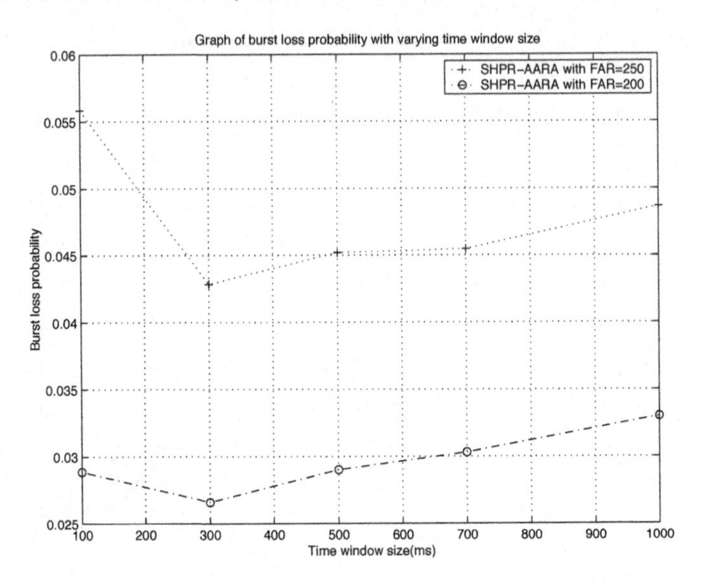

Fig. 5.9. Burst loss probability versus time window size (ms)

This is because AARA tends to prefer shorter paths when traffic load increases. The mean hop-length for the SAR algorithms is larger since SAR treats the two paths equally. Therefore on the average it uses the longer path more often than AARA since it does not adapt to the traffic load.

The impact of varying the time window size on the performance of AARA for two traffic load values, 250 $flows/sec$ and 200 $flows/sec$ is evaluated by simulations of SHPR-AARA, which is close to WSHPR-AARA and does not require prior knowledge of the traffic demands. The burst loss probability first decreases then increases with increasing time window size. The efficiency of SHPR-AARA depends on the accuracy of the traffic measurements and when the time window size is small, the collected traffic statistics reflect only the short-term traffic load conditions and may not be accurate. As a result, load balancing based on it does not work very well. Further, a small time window size results in frequent adjustments of traffic assignments, which may make the network unstable. When the time window size becomes too large, performance also starts to degrade. Here, the large window size renders the algorithm incapable of tracking dynamic changes in traffic loads and yields smoothed out traffic loads which are not reflective of the actual load conditions.

5.2.2 The streamline effect in OBS networks

An important part of a load balancing algorithm is to collect information about the level of congestion at various parts of the network. In an OBS network, congestion level at a link is indicated through burst loss probability. If the load balancing algorithm is online and a majority of network traffic is best-effort, it is sufficient to measure the burst loss probability at a link to know its congestion level. However, this is not possible if the algorithm is offline or if a majority of network traffic is reservation-based QoS traffic. In the latter case, what is important is the *future* burst loss probability when the traffic corresponding to a reservation is transmitted. In these scenarios, the load balancing algorithm must *estimate* the burst loss probability at a link.

Traditionally, the $M|M|k|k$ queueing model is adopted in performance evaluation of OBS networks. This model assumes that the input traffic to an OBS core node is Poisson, which is equivalent to having an infinite number of independent input streams. However, the number of input streams to a core node is bounded by the small number of input links. It makes the $M|M|k|k$ queueing model inaccurate. In [12, 13], a new and more accurate analytical model[4] is proposed to estimate the burst loss probability at a link for OBS networks. It takes into account a newly observed phenomenon called the *streamline effect*.

Consider an OBS core node with a number of input links connected to an output link, each with two wavelengths, as shown in Figure 5.10. To facilitate discussion, a *stream* is defined as the aggregate of all burst flows arriving in a common input link and destined for a common output link. The streamline effect is the phenomenon wherein bursts travelling in a common link are streamlined and do not contend with each other until they diverge.

[4] Portions reprinted, with permission, from (M. H. Phùng, D. Shan, K. C. Chua, and G. Mohan, "Performance Analysis of a Bufferless OBS Node Considering the Streamline Effect," *IEEE Communications Letters*, vol. 10, no. 4, pp. 293-295) ©[2006] IEEE.

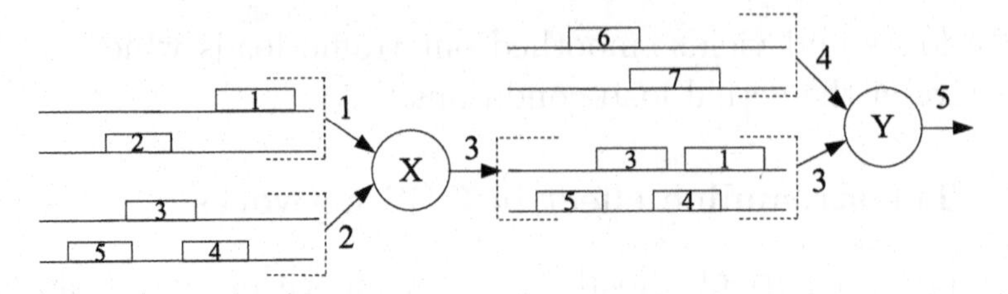

Fig. 5.10. Illustration of the streamline effect

The reason is because there is no buffer inside an OBS network. Therefore, once the contentions among them are resolved at the first link where they merge, no intra-stream contention will occur thereafter. This effect is illustrated in Figure 5.10. Burst streams 1 and 2 merge at node X. After any burst loss that might happen at node X, the remaining bursts are streamlined in output stream 3 and no further contention will happen among them. However, they may still experience contention with other burst streams that merge at downstream nodes.

The significance of this streamline effect is two folds. Firstly, since bursts within an input stream only contend with those from other input streams but not among themselves, their loss probability is lower than that obtained from the M/M/k/k model. This has major implications for traffic engineering algorithms that need to predict burst loss probability at a link. Secondly, the burst loss probability is not uniform among the input streams. The higher the burst rate of the input stream, the lower its loss probability. Therefore, if traffic within an OBS network is encouraged to form major flows with fewer merging points, the overall loss rate will be reduced.

The performance analysis is as follows. Consider two systems **A** and **B** as shown in Figure 5.11. Each node in the figure has W wavelengths per link with full wavelength conversion capability and no FDL buffer. The two systems receive identical input traffic with rate λ. The input traffic to system **A** is split into N streams with rates $\lambda_1, \ldots, \lambda_N$ coming into nodes 1,…,N. That is, $\lambda = \sum_{i=1}^{N} \lambda_i$. On the other hand, the entire input traffic λ is fed into node B in system **B**. To determine the burst loss probability of each input stream to node A, it is required to prove that the two

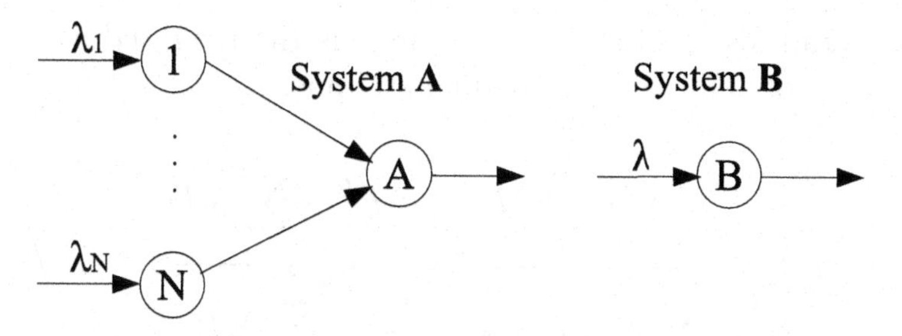

Fig. 5.11. Two equivalent systems

systems are equivalent in terms of burst loss. That is, for any burst that is lost in one system, there is a corresponding burst lost in the other.

Let M_i, M_A and M_B be the number of bursts arriving simultaneously at node i ($1 \leq i \leq N$), node A and node B, respectively. For a burst to be dropped at a node, the number of overlapping bursts must be greater than W.

- Consider a burst lost in system **A**. There are two places it can occur, either at a node i ($i \in [1, \ldots, N]$) or node A. If the burst is dropped at node i, $M_i > W$. Alternatively, if it is dropped at node A, $M_A > W$. Since M_B is greater than both M_i and M_A, $M_B > W$, i.e., there is a burst dropped at node B of system **B**.
- Consider a burst lost in system **B** or $M_B > W$. Since systems **A** and **B** receive identical input traffic, $M_B = \sum_{i=1}^{N} M_i > W$. There are two ways in which this inequality can happen. First, one particular term $M_i > W$, which implies that a burst is dropped at node i in system **A**. Alternatively, no term in the sum is larger than W, which implies that no burst is dropped at any of the nodes 1,. . . ,N. So $M_A = \sum_{i=1}^{N} M_i > W$, i.e., there is a burst dropped at node A in system **A**.

Having established that the two systems are equivalent, the loss probability of input stream i to node A can be calculated by equating the number of lost bursts for that stream in both systems. Let P_i, P_i^A and P^B be the loss probabilities of stream i at node i, node A and node B, respectively. Then

$$\lambda_i P^B = \lambda_i P_i + \lambda_i (1 - P_i) P_i^A \Rightarrow P_i^A = \frac{P^B - P_i}{1 - P_i}. \qquad (5.11)$$

The overall loss probability at node A is the weighted average of the loss probabilities of individual streams.

$$P^A = \frac{\sum_{i=1}^{N} \lambda_i (1 - P_i) P_i^A}{\sum_{j=1}^{N} \lambda_j (1 - P_j)} = \frac{\lambda P^B - \sum_{i=1}^{N} \lambda_i P_i}{\lambda - \sum_{j=1}^{N} \lambda_j P_j}. \tag{5.12}$$

If the input streams to node i ($1 \leq i \leq N$) and node B have Poisson arrival distribution, which is a reasonable assumption in backbone networks according to [16], P_i and P^B can be determined from the Erlang B formula with W wavelengths and total load ρ as follows

$$P = B(\rho, W) = \frac{\rho^W / W!}{\sum_{n=0}^{W} \rho^n / n!}. \tag{5.13}$$

It is also noted that in a real network, true Poisson traffic like the inputs to nodes $1, \ldots, N$ and node B does not exist and neither do λ and λ_i. Nevertheless, λ and λ_i can be derived from the input rates to node A using (5.13).

The above results hold well in the limiting cases. If $N = 1$, $P_1 = P^B \Rightarrow P_1^A = P^A = 0$. On the other hand, if $N \to \infty$, $\lambda_i \to 0 \Rightarrow P_i = 0$. Thus, $P_i^A = P^A = P^B$. In this second special case, the assumption of the M/M/k/k model is valid and it produces the same result as the above model. However, it should be noted that N is bounded by the number of input links to a node, which is often very small. Therefore, the second limiting case does not have practical significance.

In [12, 13], a simulation is developed to verify the accuracy of the streamline model. The measured loss probabilities of input streams at node A and those given by the $M|M|k|k$ model and the streamline model are plotted against various parameters. Apart from validating the streamline analysis, an additional goal is to determine which parameters affect the accuracy of the $M|M|k|k$ model. This information will be useful in determining the scenarios under which the Erlang B formula can be used in OBS networks. In the simulations, the same topology as system **A** in Figure 5.11

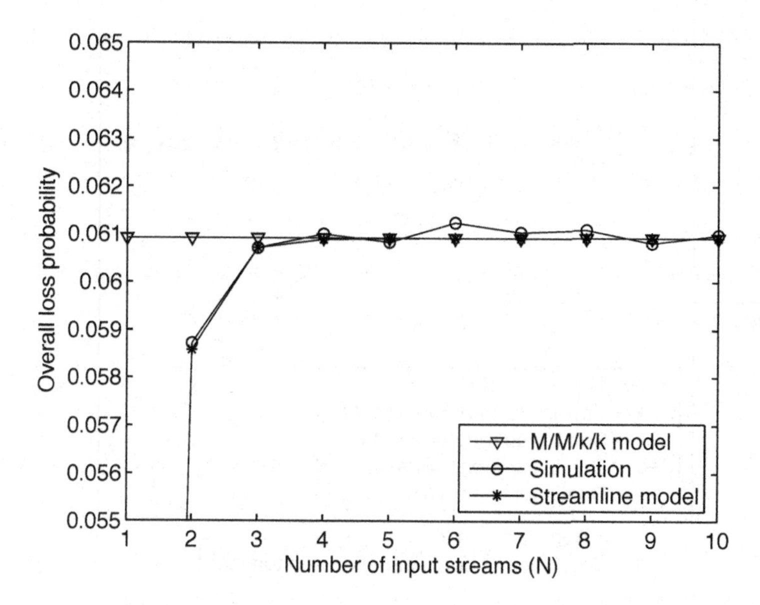

Fig. 5.12. Overall burst loss probability versus number of input streams

is used. Unless otherwise stated, each link has 8 wavelengths and each wavelength has a capacity of 40 Gbps. In the simulations, the input burst streams have Poisson arrival pattern and the burst lengths are uniformly distributed between 50 and 51.5 KB. The burst length distribution is chosen to emulate the output produced by a size-limited burst assembly algorithm such as in [10].

Figure 5.12 shows the overall loss probability versus the number of input streams N. The total offered load at the output link of node A is 0.6. All input streams have equal rates. The results show that the streamline model always gives accurate results. The $M|M|k|k$ model, on the other hand, is only accurate when $N \geq 3$. There is a slight deviation when $N = 2$, which is tolerable. However, when $N = 1$, the measured loss probability falls to zero but the $M|M|k|k$ model still gives the same result. This special case is predicted by the analysis.

In another experiment, the number of input streams is kept constant at 2 and the traffic contribution of each stream is varied. The results in Figure 5.13 show very different curves for the loss probabilities of the two streams. For the smaller stream, the loss

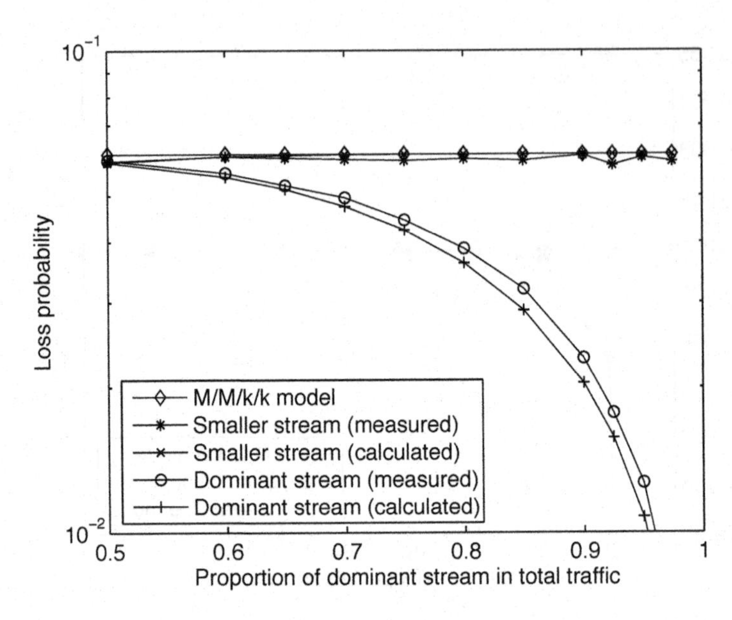

Fig. 5.13. Loss probabilities of individual streams versus traffic proportion of dominant stream

probability is relatively constant. This is expected from the analysis since for the small stream, P_i is very small and hence, $P_A^i \approx P_B$. For the dominant stream, its loss probability falls rapidly as its traffic share gets closer to unity. This can also be explained from the analysis since $P_i \rightarrow P_B$ as the dominant flow gets larger. Thus, $P_A^i \rightarrow 0$. However, this result cannot be predicted by the $M|M|k|k$ model, which gives a constant loss probability.

The results in Figure 5.13 can also be explained intuitively by recalling that the bursts within one input stream do not contend with each other and only contend with those from other streams. For the dominant stream, their bursts have few competitors from the small stream and hence their small loss probability. On the other hand, the bursts from the small stream have almost as many competitors from the dominant streams as in the case of a large number of input streams. Therefore, their loss probability is close to that given by the $M|M|k|k$ model.

The effect of the number of wavelengths per link on the accuracy of the models is also examined. There are two input streams

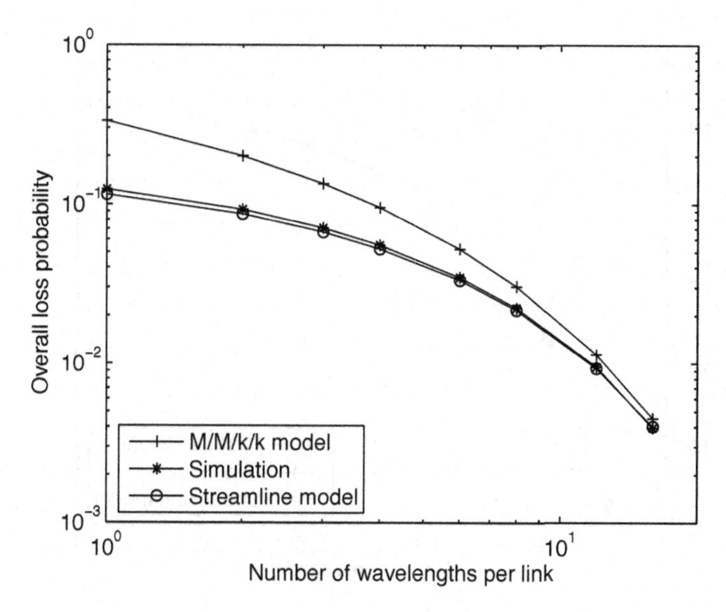

Fig. 5.14. Overall loss probability versus number of wavelengths per link

with rates of 0.4 and 0.1 to the node. As shown in Figure 5.14, the streamline model always gives accurate results. On the other hand, the $M|M|k|k$ model is only accurate when the number of wavelength is large.

The overall loss probability versus the total offered load is plotted in Figure 5.15. The number of input streams is set at 2 and both streams have equal rates. The analysis and the $M|M|k|k$ model both give results that closely match the simulation results at all offered loads. In other words, the total offered load does not play a role in the accuracy of the $M|M|k|k$ model.

5.2.3 Load balancing for reservation-based QoS traffic

In [12], a load balancing algorithm[5] for reservation-based QoS traffic is proposed. The algorithm is developed specifically for the e2e absolute QoS framework in [3, 4, 17]. When a new LSP needs to be

[5] Portions reprinted, with permission, from (M. H. Phùng, K. C. Chua, G. Mohan, M. Motani, and T. C. Wong, "The Streamline Effect in OBS Networks and Its Application in Load Balancing," in *Proc. 2nd International Conference on Broadband Networks*, pp. 283–290) ©[2005] IEEE.

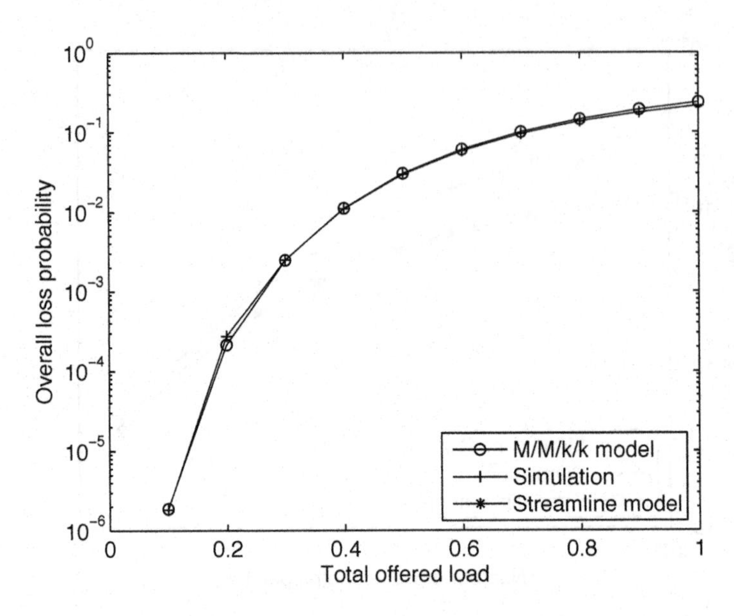

Fig. 5.15. Overall loss probability versus total offered load

established, the absolute QoS framework reserves resources along the path of the LSP to satisfy the LSP's request or denies the request if insufficient resource is available. However, the framework only operates over a predetermined path for each LSP. The load balancing algorithm extends it to consider multiple paths between an ingress/egress node pair and chooses a path based on the calculated cost of the paths. Since the future burst loss probability of a link if/when a request is accepted cannot be measured, the scheme needs to estimate it based on the current offered load to the link and the amount of bandwidth the new LSP requests. The streamline model is used for this purpose to improve its performance.

The implication of the streamline effect to the absolute QoS framework is two folds. The first obvious and beneficial implication is that a more accurate loss probability formula will enable tighter estimates and better admission control decisions for the admission control routine at each core node. The second implication can be understood by looking at Figure 5.13. It shows that the loss probabilities of dominant and non-dominant streams are vastly different if the traffic share of the dominant stream is large. Therefore, there are some issues in guaranteeing thresholds for individual classes at a core node. These can be divided into the following two categories.

- *Inter-class contention:* This is the case where the traffic classes in the dominant stream are different from those in the small streams. There is no problem in this scenario since the absolute QoS framework uses preemptive differentiation to resolve contention among classes. The preemptive differentiation will override the streamline effect.
- *Intra-class contention:* This is the case where one or more traffic classes exist in both the dominant stream and the small streams. Within such classes, the traffic portion from the dominant stream will experience lower loss probability than those from other streams. This is not desirable since some LSPs in the small streams may experience loss probability over the guaranteed threshold of their class. There are two solutions for this. The first solution is to build the streamline effect into the preemption mechanism to compensate for it, i.e., to mark the bursts from the dominant stream and over-drop them. The second solution is to divide the affected class into two subclasses: one for the traffic portion from the dominant stream and one for the rest. In effect, it converts intra-class contentions into inter-class contentions which the preemptive differentiation mechanism can handle. This is preferred since it does not introduce additional complexity into the preemption mechanism.

The dynamic route selection algorithm for the absolute QoS framework is described below. It is similar to that in [9, 8]. However, there are differences due to the fact that absolute QoS traffic is involved and the streamline effect is considered.

At an ingress node, two link-disjoint paths are determined in advance for each egress node. One of the paths is the shortest path. At any time, a cost is associated with each of the paths, which is the sum of the costs of the intermediate links along the path. For the longer path, a penalty term is also added to its cost. Let Δh

be the hop difference between the two paths. The penalty term is given by $PA = \Delta h \times \tau$ where τ is a system control parameter.

When a new LSP needs to be set up, the less costly path will be selected to initiate the QoS reservation process. Since path cost is calculated based on the loss probability on that path, the less costly path will offer a better chance of acceptance. However, it is possible that changes in traffic conditions at the intermediate links have not been updated in the path costs. Therefore, when the selected path does not accept the reservation request, the reservation process is started again on the other path. During the reservation process, the costs of the intermediate links of the path involved are also collected. This is signalled back to the ingress node using the acknowledgement message to update the cost of the path involved. If there is no reservation process on a path within a time period T (e.g., because its cost is too high and therefore it is not selected), a probe packet is sent out to update its cost. Note that existing LSPs are not shifted between the two paths. This is to avoid out-of-order burst arrivals and the synchronisation problem, which is caused by the feedback between traffic shifting and path costs. The number of LSPs in backbone networks should be sufficiently high to effectively balance traffic between the paths.

For each output link at a core node, the node needs to keep track of the traffic contribution λ_i from each input link in addition to the total traffic λ at the output link. When a reservation request with bandwidth b_0 arrives from input link i, the node substitutes λ_i with $\lambda_i' = \lambda_i + b_0$ and calculates the estimated loss probability of traffic from that input link using Equation (5.11). This is recorded in the reservation message as the link cost. On the other hand, if a probe packet arrives, no substitution of λ_i is necessary.

The use of Equation (5.12) in calculating the link cost differentiates reservation requests from dominant and non-dominant streams. That is, it gives lower costs to requests coming from the dominant stream. This will encourage the dominant stream to grow, which in turn lowers its cost. Thus, the proposed algorithm does not only balance but also judiciously redistributes traffic in an OBS network to further reduce burst loss.

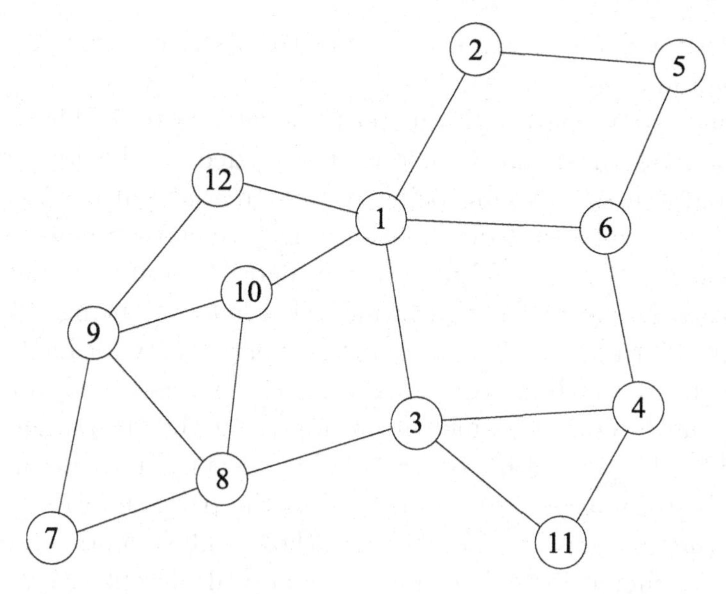

Fig. 5.16. Random network topology with 12 nodes

Performance evaluation

A simulation model is used to investigate the performance of the proposed load balancing algorithm for absolute QoS traffic. A randomly generated network topology as shown in Figure 5.16 is used. The network has 12 nodes, 18 bidirectional links and 8 data wavelengths per link. The capacity of a wavelength is 10 Gbps. Full wavelength conversion and no FDL buffer are assumed at all nodes. The LAUC-VF scheduling algorithm presented in [10] is used to schedule bursts. The bursts are generated from a size-limited assembly algorithm with the size limit at 50 kB. Thus, the mean burst length is about 10 μs.

Six ingress/egress node pairs are selected to transmit in the simulation, namely (7;5), (7;11), (9;5), (9;11), (12;5), (12;11). There are two paths between each node pair. One path is computed using a shortest path routing algorithm, while the other is the link-disjoint next-shortest path. New LSPs arrive at each ingress node according to a Poisson process. For simplicity, all LSPs are assumed to have the same burst rate of 4000 bursts/s. Hence, it takes 200 concurrently active LSPs to have 100% offered load. A new LSP will request an e2e loss probability of either 0.02 or 0.05 at random.

The system control parameter is set at $\tau = 0.01$. This is used to calculate the penalty cost that is added to the longer path to avoid excessive traffic being routed to that path, which may result in performance degradation due to increased consumption of network resources. Three routing algorithms are simulated for comparison: shortest path routing (SP); load balancing using the Erlang B formula to estimate burst loss probability for LSP admission control and link cost calculation (LB-E) and load balancing using the formula obtained from analysing the streamline effect (LB-S). The percentage of LSPs being accepted into the network over the total arriving LSPs is used as the performance metric.

In the first experiment, the case wherein all six node pairs have identical offered loads is examined. The LSP acceptance percentage of the three schemes is plotted against the offered load per node pair in Figure 5.17. The results show that LB-S has the best performance followed by LB-E and SP. Since SP always chooses the shortest path to use, congestion may easily develop at links where some shortest paths join. On the other hand, LB-S and LB-E may spread traffic to longer paths and hence perform better. Between the two load balancing algorithms, LB-S takes into consideration the streamline effect when calculating path costs. Therefore, major streams are encouraged to form, which reduce burst loss probability and increase LSP acceptance probability. In addition, LB-S uses the streamline formula in LSP admission control, which gives added improvement over LB-E.

The percentage improvement of LB-S and LB-E over SP is plotted against the offered load per node pair in Figure 5.18. The improvement increases as the network load increases. When the traffic load is light, congestion at bottleneck nodes is not so severe and there is not much room for improvement by load balancing. However, when the traffic load becomes heavy, the two load balancing schemes are able to mitigate congestion, explaining their increased percentage of improvement.

Figure 5.19 shows the mean hop length traversed by a burst against the offered load per node pair. The mean hop length

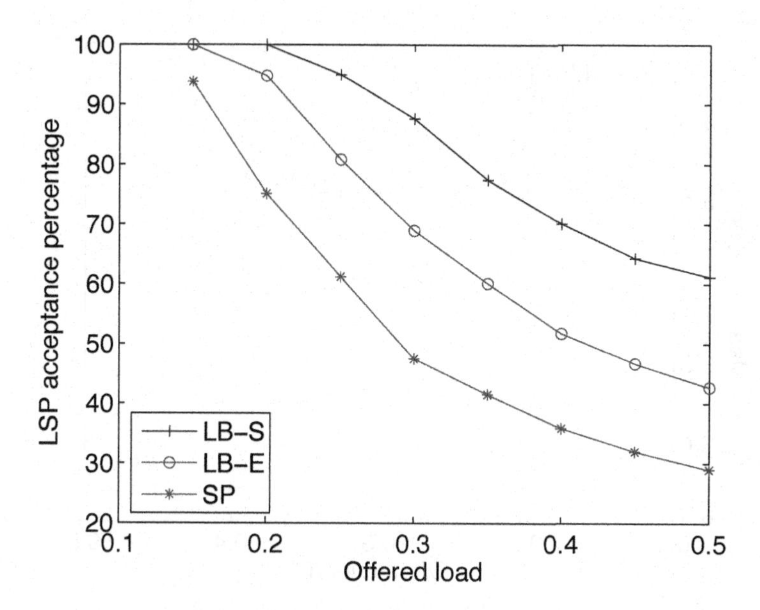

Fig. 5.17. Percentage of LSP accepted versus offered load per node pair for identical traffic demands

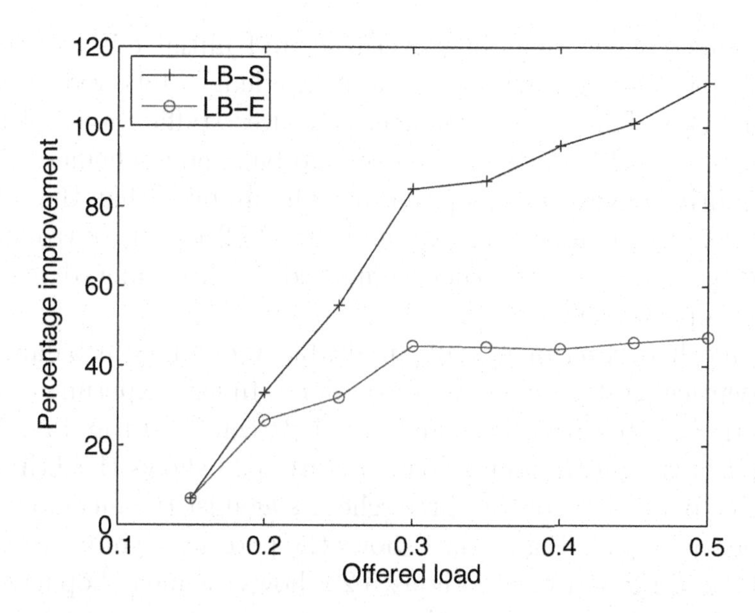

Fig. 5.18. Percentage improvement over shortest path routing for identical traffic demands

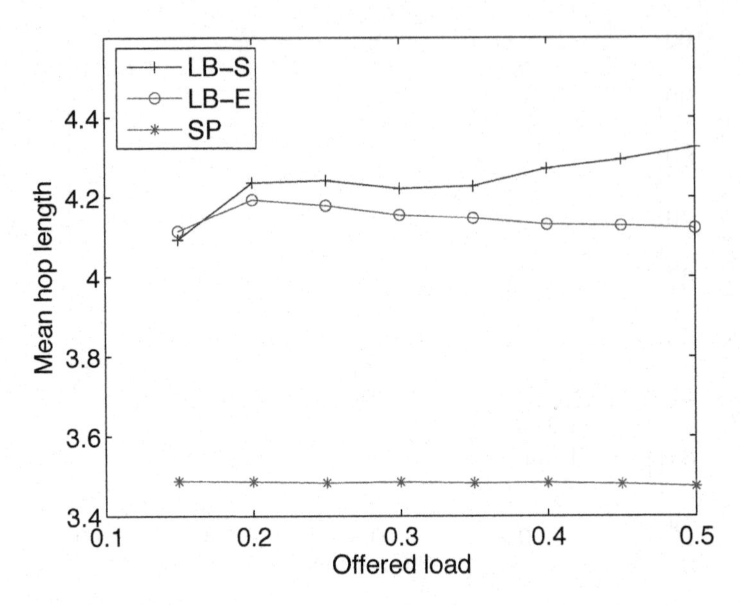

Fig. 5.19. Mean hop length versus offered load per node pair for identical traffic demands

could affect delay, signalling overhead and initial offset time in the network. SP has the lowest mean hop length followed by LB-E and LB-S. This is obvious since it routes traffic only through the shortest paths. Between the two load balancing schemes, LB-S has slightly larger mean hop length. This implies that the additional delay, signalling overhead and initial offset time introduced by LB-S are rather low when compared to the achieved performance improvement.

In another experiment, the applicability of LB-S in balancing non-identical traffic demands is examined. In this experiment, the LSP arrival rates are taken randomly between $l - 0.1$ and $l + 0.1$, where l is the mean offered load. Figure 5.20 shows the LSP acceptance percentage of the three schemes against the mean offered load per node pair. Figure 5.21 shows the percentage improvement of LB-S and LB-E over SP. Figure 5.22 shows the mean hop lengths of the three schemes. From these figures, similar observations as in the case of identical traffic demands can be made. The only difference is that in this experiment, LB-S has slightly lower mean

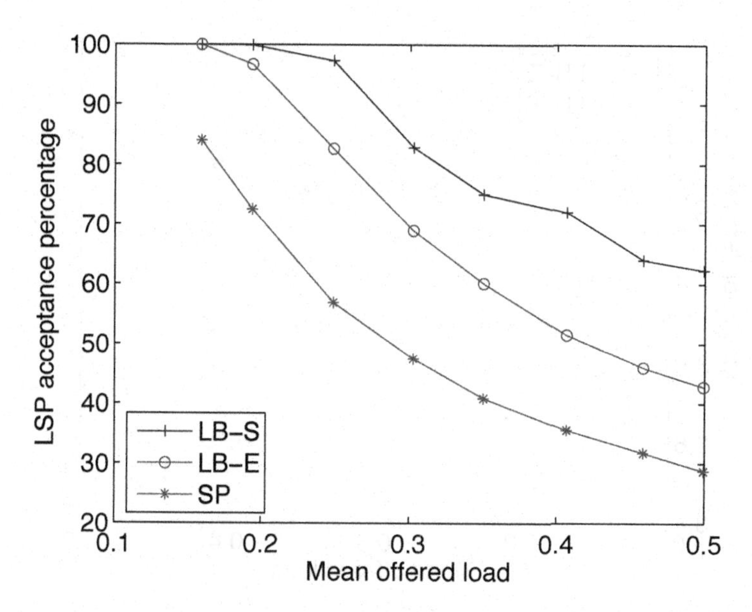

Fig. 5.20. Percentage of LSP accepted versus mean offered load per node pair for non-identical traffic demands

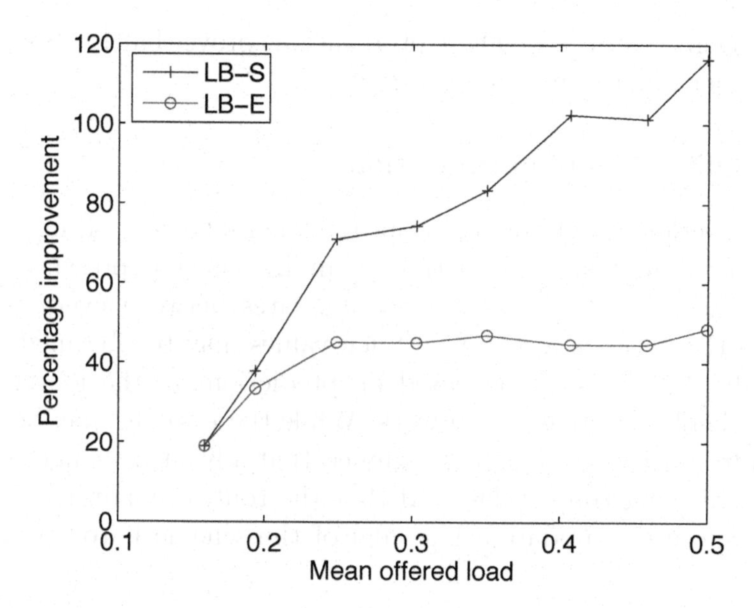

Fig. 5.21. Percentage improvement over shortest path routing for non-identical traffic demands

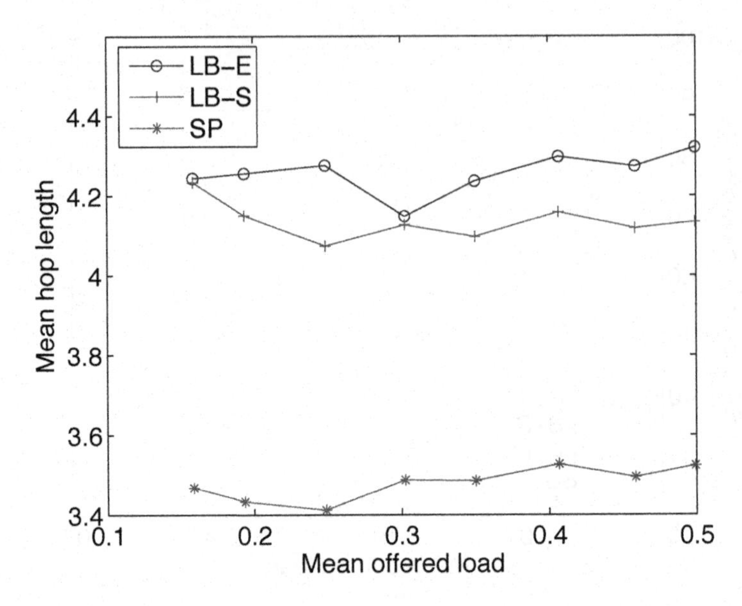

Fig. 5.22. Mean hop length versus mean offered load per node pair for non-identical traffic demands

hop lengths than LB-E. These observations prove that LB-S works well under different traffic scenarios.

5.2.4 Offline route optimisation

In this section, an offline route optimisation scheme[6] [14] for OBS networks that makes use of the streamline effect is presented. It is assumed that the long-term traffic rates between every node pair in the network are known. Such values may be obtained empirically, or they may be based on predictions of the long-term demands placed upon the network. While these values may be updated from time to time, it is assumed that any such changes take place over long time scales, and that the routes remain fixed between successive updates. The goal of the scheme is to optimise

[6] Portions reprinted, with permission, from (Q. Chen, G. Mohan, and K. C. Chua, "Offline Route Optimization Considering Streamline Effect in Optical Burst Switching Networks," in *Proc. IEEE International Conference on Communications*) ©[2006] IEEE.

traffic routes in the network so as to minimise the overall burst loss.

To solve the problem of offline route optimisation, it is necessary to estimate the burst loss at a link given an input traffic load. Offline route optimisation is initially studied in [15] using the Erlang B formula to estimate the burst loss. However, it has been noted in section 5.2.2 that the Erlang B formula can give inaccurate estimates of burst loss. Therefore, the streamline effect is considered in the scheme presented in this section in order to obtain a more accurate burst loss estimation at a link.

From (5.11) and (5.12), we have

$$P^A \sum_{j=1}^{N} \lambda_j (1 - P_j) = \sum_{i=1}^{N} \lambda_i (1 - P_i) P_i^A = \lambda P^B - \sum_{i=1}^{N} \lambda_i P_i. \quad (5.14)$$

Denote

$$G(a, W) = a \times B(a, W) = a(\frac{\frac{(Wa)^W}{W!}}{\sum_{m=0}^{W} \frac{(Wa)^m}{m!}}) \quad (5.15)$$

as the burst loss amount estimated by the Erlang B formula at a link with load a and W wavelengths, from (5.14), we have the amount of burst loss at node A as

$$G(\rho, W) - \sum_{i=1}^{N} G(\rho_i, W). \quad (5.16)$$

That is, the loss amount at a link is equal to the loss amount estimated by the Erlang B formula minus the contentions within the same individual input streams.

MILP formulation

Based on the new loss estimation formula, an MILP formulation for the offline route optimization problem is stated as follows:

Given an OBS network topology and a traffic demand, it is required to determine a route for each flow to minimize the overall burst loss.

The following notations are used:

- *links*: the set of the links in the network.
- *nodes*: the set of the nodes in the network.
- F: the set of the flows. Each flow is identified by a tuple $< s, d >$, where s and d are the source and destination node, respectively.
- W: the number of wavelengths in one link.
- $Head(v)$: the links starting from node v.
- $Tail(v)$: the links ending at node v.
- $Up(l)$: the upstream end node of link l.
- $Down(l)$: the downstream end node of link l.
- $Prev(k)$: the set of the links which ends at the node where link k starts. In other words, the links previous to link k.
- $\rho_{s,d}$: the traffic load of flow $< s, d >$.
- $x^k_{s,d}$: it is 1 if the flow $< s, d >$ goes through link k. Otherwise 0.
- $z^{l,k}_{s,d}$: it is 1 if flow $< s, d >$ goes through the concatenation of link l and k. Otherwise 0. The variable exists only if $l \in prev(k)$.
- ρ^k: the load over link k.
- $\theta^{l,k}$: the load over the concatenation of link l and k. The variable exists only if $l \in prev(k)$.
- δ: a small value (set to 10^{-8} in the numerical evaluation). It keeps the link cost greater than zero and prevents a loop in the route found.
- $Loss^k$: the burst loss over link k.

Problem Formulation:
Objective : Minimize $\sum_{k \in links} Loss^k$
Subject to:
1. Flow conservation constraints:

$$\sum_{k \in tail(v)} x^k_{s,d} - \sum_{k \in head(v)} x^k_{s,d} = \begin{cases} 1 & \text{if } v = s \\ -1 & \text{if } v = d \\ 0 & otherwise \end{cases} \qquad (5.17)$$

2. By the definition of $z^{l,k}_{s,d}$, it can be formulated as $z^{l,k}_{s,d} = x^k_{s,d} \times x^l_{s,d}$. However, such a formulation makes this problem non-linear. Therefore, the following linear constraints are defined. They give the same results as the multiplication equation as long as all the variables involved are boolean.

$$z^{l,k}_{s,d} \leq (x^k_{s,d} + x^l_{s,d})/2 \tag{5.18}$$

$$z^{l,k}_{s,d} \geq (x^k_{s,d} + x^l_{s,d})/2 - 0.5 \tag{5.19}$$

$$\forall < s, d > \in F, \forall k \in links, \forall l \in prev(k)$$

3. The definition of $\rho^k, \theta^{l,k}$ and $Loss^k$

$$\rho^k = \sum_{<s,d>\in F} \rho_{s,d} \times x^k_{s,d} \qquad \forall k \in links \tag{5.20}$$

$$\theta^{l,k} = \sum_{<s,d>\in F} \rho_{s,d} \times z^{l,k}_{s,d} \qquad \forall k \in links, \forall l \in prev(k) \tag{5.21}$$

$$Loss^k = \hat{G}(\rho^k, W) - \sum_{l\in prev(k)} \hat{G}(\theta^{l,k}, W) + \delta \qquad \forall k \in links \tag{5.22}$$

Because $G(\rho, W)$ is a non-linear function, a piecewise linear function is used to approximate it to make the formulation linear. Thus, $\hat{G}(\rho, W)$ is a piecewise linear function to approximate $G(\rho, W)$ with interpolation.

Heuristic Algorithm

Since the computation to solve an MILP problem is usually intensive, a heuristic algorithm, Streamline Effect Based Route Layout Heuristic (SL-Heur), is proposed in [14]. SL-Heur is composed of two phases. The first phase performs initialization, where each flow is assigned the shortest path. The second phase adopts iterative techniques to optimize the route layout. In each iteration, some flows are randomly chosen to have their routes re-computed. If the new routes help reduce the overall loss, these flows will have their routes updated. Specifically, SL-Heur works as follows:

1. Each flow is assigned the shortest path by Dijkstra's algorithm. The cost of a link is the loss brought by the new flow according to the loss estimation formula proposed. The details of the link cost definition are given later. After the path of a flow is decided, update the load of the links this flow traverses.
2. Randomly select M flows and remove their traffic load in the network. Then use Dijkstra's algorithm as in step 1 to find a route for each of the M flows. If the re-routing reduces the total loss, update the routes for these M flows. Otherwise keep the original routes.
3. Repeat step 3 until the stopping criterion is met.

The link cost of link from i to j, $cost(i,j)$, is computed as follows:

$$new = G(\rho_E^{i,j} + \rho', W) - G(\theta_E^{P(i),i,j} + \rho', W) \tag{5.23}$$

$$old = G(\rho_E^{i,j}, W) - G(\theta_E^{P(i),i,j}, W)$$

$$cost(i,j) = new - old + \delta \tag{5.24}$$

The notations used above are:

- ρ' : the traffic load of the flow whose route is to be determined.
- $\rho_E^{i,j}$: the existing traffic load over link $< i,j >$ before the new flow is deployed.
- $\theta_E^{p,i,j}$: the existing traffic load going through both link $< p,i >$ and $< i,j >$ before the new flow is deployed.
- $P(i)$: the previous node of node i in the shortest path from the source node to node i. Note that in Dijkstra's Shortest Path First algorithm, when link $< i,j >$ is considered as a possible next hop in the shortest path, the shortest path to node i has been determined.
- *new* and *old* : The difference between *new* and *old* is the loss increase on the link if the new flow is introduced. The burst losses over link $< i,j >$ with and without the flow going through are $G(\rho_E^{i,j} + \rho', W) - G(\theta_E^{P(i),i,j} + \rho', W) - \sum_{p \neq P(i)} (\theta_E^{p,i,j}, W)$ and $G(\rho_E^{i,j}, W) - G(\theta_E^{P(i),i,j}, W) - \sum_{p \neq P(i)} (\theta_E^{p,i,j}, W)$, respectively. Since

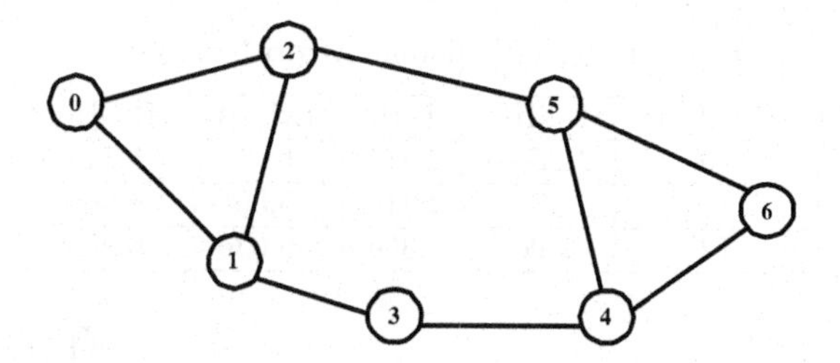

Fig. 5.23. A 7-node network topology

we are concerned with the loss difference between these two values, we remove the common item of $-\sum_{p \neq P(i)} (\theta_E^{p,i,j}, W)$ and have the expression of *new* and *old* as given above.
- δ: a small value (It is set to 10^{-8} in the numerical experiment). It keeps the link cost greater than zero and prevents a loop in the route found.

Numerical Evaluation

Simulation is used to evaluate the effectiveness of the MILP solution (referred as SL-MILP) and SL-Heur presented above. For comparison, the MILP and the heuristic presented in [15](referred as Erl-MILP and Erl-Heur), where the streamline effect is not considered and Erlang B formula is employed to estimate the loss, are also implemented. The experiments are conducted over two network topologies. One is a random 7-node topology, the other one is the 14-node NSFNET topology. There are 32 wavelengths on each link.

For the 7-node topology shown in Figure 5.23, the algorithms, SL-MILP, SL-Heur, Erl-MILP and Erl-Heur, try to find the optimal route layout for three traffic scenarios, where 10, 11 and 12 node pairs are randomly chosen to be active respectively. The traffic load between each node pair is 0.2. The results are listed in Table 5.1.

For the 14-node NSFNET topology shown in Figure 5.24, only the two heuristics, SL-Heur and Erl-Heur, are evaluated. The two MILP algorithms are computationally intensive for this topology and are therefore ignored. There exists a flow between each node

Table 5.1. Burst Loss Rates of Different Algorithms in the 7-Node Network

No. of Flows	SL-MILP	SL-Heur	Erl-MILP	Erl-Heur
10	1.25e-6	2.21e-6	4.87e-4	5.11e-4
11	2.13e-6	2.41e-6	5.56e-4	5.66e-4
12	4.30e-6	4.38e-6	6.03e-4	6.25e-4

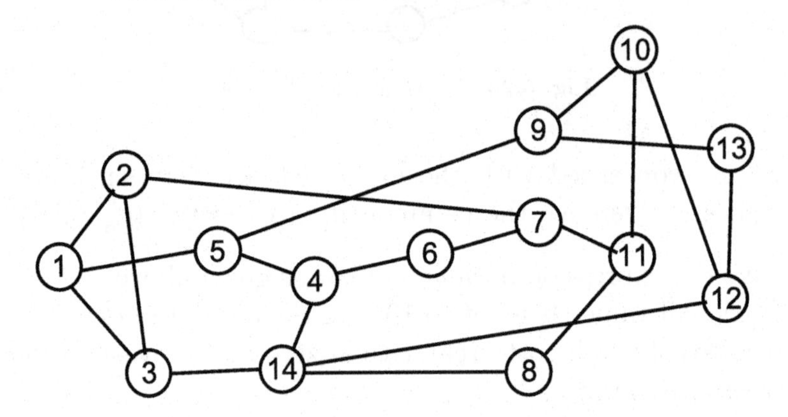

Fig. 5.24. 14-node NSFNET topology

pair. Two traffic load distribution patterns are tested. In the first pattern, the uniform pattern, traffic load of each flow is equal, while in the second, the non-uniform pattern, the traffic load of each flow is randomly chosen from an interval. The experimental results are given in Figure 5.25.

As the results show, the route layouts optimized by the streamline-based algorithms, SL-MILP and SL-Heur, result in lower overall burst loss than those by those based on Erlang B formula. When the traffic load is light, the advantage is greater, and the advantage decreases as the load becomes higher.

The observation of the advantage reduction under higher load is consistent with earlier results in [15]. As the load becomes higher, the burst loss difference between the route layout found by the traffic engineering algorithm and that given by the SPF (Shortest Path First) layout is remarkably reduced. The reason of this phenomenon may be that, when the load is higher, there are less choices in the route selection for the flow deployment if all the flows are admitted. As a result, the performance of different route selection algorithms are closer.

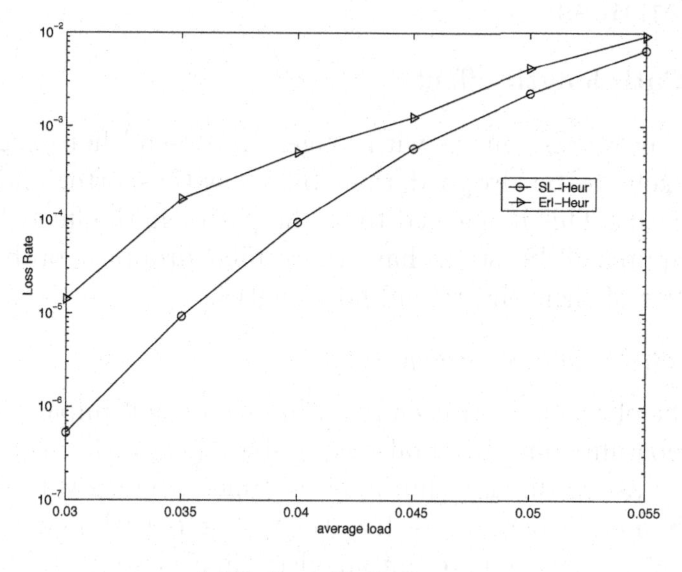

(a) Uniform Traffic Pattern

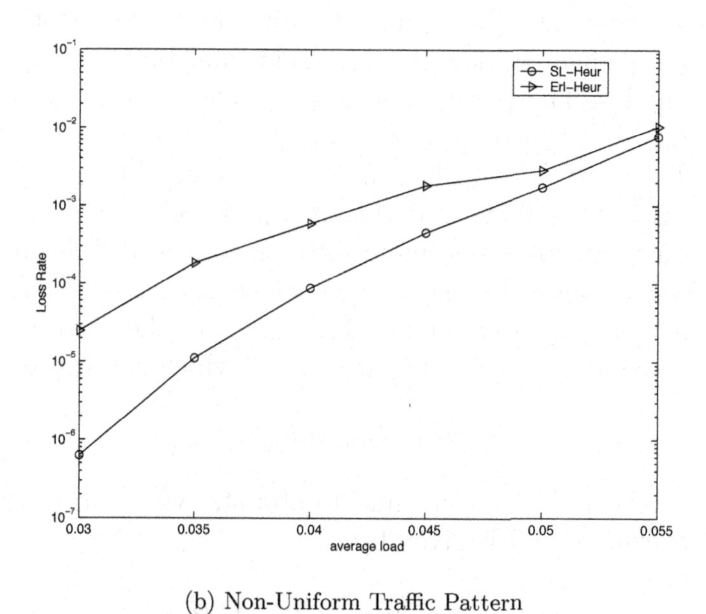

(b) Non-Uniform Traffic Pattern

Fig. 5.25. Burst Loss Rates of Different Algorithms for NSFNET topology

5.3 Fairness

5.3.1 Path length effect

In OBS networks, bursts with longer end-to-end hop lengths are more likely to be dropped than bursts with shorter end-to-end hop lengths. This is referred to as the path length effect. Fairness here requires all bursts to have the similar dropping probabilities regardless of their end-to-end hop lengths.

Initial offset time-based algorithm

Since the offset time for each burst has significant impact on burst loss probability, an initial offset time-based algorithm is proposed in [18] to assign different initial offset time to bursts with different number of end-to-end hops. Intuitively, bursts with longer end-to-end path will have a larger initial offset time as compared to bursts with shorter end-to-end path to reduce the path length effect. In this algorithm, there is an upper limit on the initial offset time to reduce queueing delay at edge nodes and buffer requirements at edge nodes. The permissible offset range is given by $[0, T_{\max}]$, where $T_{\max}$ is given by $[h_{\max}\delta + O_{\max}L]$, in which $h_{\max}$ is the diameter for a network which is the maximum number of hops for all ingress and egress pairs, δ is the configuration delay at each hop, $O_{\max}$ is a predefined maximum offset factor, and L is the mean burst length. Let n be the number of service classes and class i has higher priority over class $i-1, i-2, \cdots, 1$. The offset is equally divided into $nh_{\max}$ segments with size B which is given by

$$B = T_{\max}/nh_{\max} \tag{5.25}$$

Then, the initial offset time for bursts with end-to-end hop length h from class i is given by

$$O_i = (i-1)\,h_{\max} + (h-1)\,B + h\delta \tag{5.26}$$

Since the end-to-end hop length is taken into consideration for the setting of the initial offset time, the path length effect can be reduced to a certain degree. However, this approach cannot adapt well to dynamic changing network states. Another scheme is proposed to reduce the path length effect by considering the network state, which is presented in the next section.

Link scheduling state information based offset management scheme

In [19][7], a link scheduling state information based offset management scheme is used to improve fairness in OBS networks. The basic idea is to choose different offset times for bursts with different e2e hop lengths to compensate for the path length effect since offset time has a significant effect on burst dropping probability during a burst contention. Particularly, each core node collects link state information that includes the dropping probabilities for bursts with different offset times. This link state information is then advertised to each source node for reference. Before deciding on the offset time for an outgoing burst, a reference e2e burst dropping probability is chosen. In [19], the burst dropping probability of a monitored one-hop burst originating from the same source node is chosen as the reference burst dropping probability. The offset time for the outgoing burst is selected to result in a similar dropping probability as the reference burst dropping probability. The e2e burst dropping probability is approximated by multiplying the individual burst dropping probabilities assuming independence of burst loss at each node. Since the source node has the latest information on the relationship between the burst dropping probability and offset time at each core node, it is possible for the source node to select an appropriate offset time to achieve the desired e2e loss probability. Since the source node always has the latest information on the link states, the scheme can adapt to dynamic traffic patterns.

Figure 5.26 shows a simple example of a 2-hop path with 3 nodes. The latest link scheduling state information is shown in Tables 5.2 and 5.3. The burst dropping probability for a burst with only 1 hop is set as the reference dropping probability. In

[7] Portions reprinted from Computer Networks, Vol. 45, S. K. Tan, G. Mohan, and K. C. Chua, Link Scheduling State Information Based Offset Management in WDM Optical Burst Switching Networks, pp 819-834, © (2004), with permission from Elsevier.

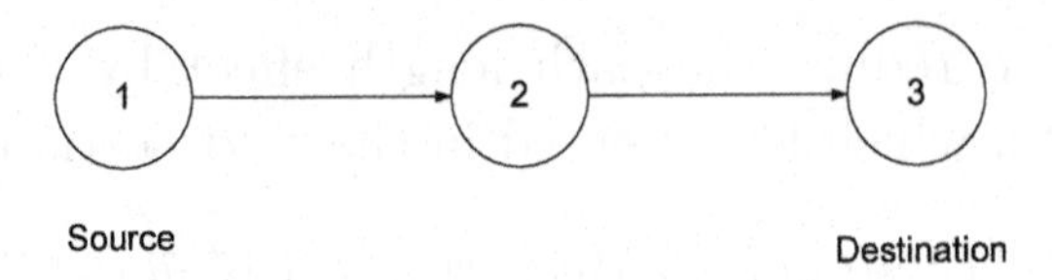

Fig. 5.26. A path with 2 hops

Table 5.2. Link state information at node 1

Offset at node 1	Link scheduling probabilities
2	0.9681
3	0.9783
4	0.9805
5	0.9824
6	0.9837
7	0.9846
8	0.9850
9	0.9902

this example, 0.9681 is set as the reference loss probability. Let the processing delay for the header packet be 2 time units. Therefore, the minimum offset time for a burst with 2 hops is 4 time units. According to the link scheduling state information of Tables 5.2 and 5.3, path loss probabilities are calculated under different initial offset times for an outgoing burst as shown in Table 5.4. The path dropping probability assuming an initial offset time is then calculated by multiplying the link scheduling probabilities (one minus the loss probability at each hop under different offset time). For example, the path scheduling probability for a burst with initial offset time of 6 units is obtained by 0.9876*0.9757, where 0.9757 is the scheduling probability at the second hop under an offset time of 4 units (initial offset time of 6 units minus 2 time units at the second hop). According to Table 5.4, a burst with an initial offset time of 8 units can achieve a scheduling probability close to the reference probability.

Search space-based algorithm

A simple search space based algorithm is proposed in [20] to alleviate the path length effect by increasing the wavelength search space for a burst gradually when it travels more hops. That is, the

Table 5.3. Link state information at node 2

Offset at node 2	Link scheduling probabilities
2	0.9581
3	0.9634
4	0.9757
5	0.9770
6	0.9779
7	0.9860
8	0.9879
9	0.9891

Table 5.4. Path scheduling probabilities under different initial offset time

Offset at node 1	Offset at node 2	Path scheduling probability
4	2	0.9394
5	3	0.9464
6	4	0.9598
7	5	0.9620
8	6	0.9632
9	7	0.9763

search space for a burst is determined by the number of hops it has passed. Specifically, let $h_{\max}$ denote the diameter of a network which is the maximum number of hops for all ingress and egress pairs. In this proposed algorithm, each burst is associated with a search space which consists of two parts. The first part is the base search space which is the same for all the bursts to avoid the problem of service starvation for bursts with small number of end-to-end hops. The second part is the adjustable search space which is associated with the number of hops that a burst has passed. The higher the number of hops a burst has passed, the larger the adjustable search space for that burst. In particular, for a burst at its i-th hop, the search space n_i is given by

$$n_i = (1 - g) W + giW/h_{\max} \tag{5.27}$$

where $g \in [0, 1]$ is a controllable parameter, W is the total number of wavelengths. It can be seen from Eq. (5.27) that $(1 - g) W$ is the base part for the search space which is independent of the number of hops and $giW/h_{\max}$ is the adjustable part which is dependent on the number of hops that a burst has passed. It can be seen that Eq. (5.27) is similar to Eq. (3.10) except that in the former, i denotes the priority level and in the latter, i denotes the number of hops.

Proactive dropping-based algorithm

A proactive dropping-based algorithm is proposed in [20] to reduce the path length effect. This RED-like algorithm proactively drops incoming bursts when the queue length of an edge node exceeds a given threshold. Burst are proactively dropped with probabilities determined by their path lengths. By setting different dropping probability for bursts according to its hop count, the path length effect can be reduced.

5.3.2 Max-min fairness

Chapters 3 and 4 have described various algorithms and mechanisms to achieve differentiated QoS in OBS networks. However, none has studied how to provide service isolation and protection among connections[8] between different source and destination pairs sharing a common link. Without service isolation and protection, misbehaving connections will send too much traffic to the core network, resulting in well behaving connections experiencing high burst dropping probabilities. A simple solution is to have each connection reserve a fixed amount of bandwidth in the core nodes and the input rate of each connection be limited by a traffic regulator at its ingress node. However, this is inefficient since unused bandwidth of some connections will be wasted. Fairness here refers to allocating bandwidth fairly among the connections such that during a congestion, well behaving connections are protected from misbehaving connections, while making full use of the available bandwidth at all times.

Fair bandwidth allocation has been extensively studied in IP packet switching networks. A well known fairness criterion is max-min fairness bandwidth allocation [22]. However, there are some

[8] A connection can be the traffic using a label switching path (LSP) between different source and destination pairs if MultiProtocol Label Switching (MPLS) is deployed for burst switching in OBS networks.

key differences between IP packet switching networks and OBS networks which make it difficult to achieve the same max-min fair bandwidth allocation in the latter. One major difference is that there is no or limited buffer in OBS networks.

Max-min fairness in OBS networks

In IP packet switching networks, a bandwidth allocation is said to be max-min fair if the allocation has a maximum minimum bandwidth allocation among all other allocation schemes and no connection can be allocated more than its request [22]. The basic principle of max-min fairness in IP packet switching networks is to make sure that each source has the same right for a resource. Thus the fair share (bandwidth) guaranteed to the well behaving connections with smaller arrival rates will be isolated from the misbehaving connections with higher arrival rates.

For OBS networks, the max-min fairness criterion is defined as follows. The loss probability of each connection will be maintained around a theoretical loss level which is determined by its fair share in a max-min fair manner. The theoretical loss level is determined such that:

i). when the fair share of one connection increases to r, the loss probabilities of those connections with smaller fair shares than r will not increase;

ii). when the fair share of one connection decreases to r, its surplus bandwidth will be fairly allocated to other connections with larger fair shares than r.

Max-min fair bandwidth allocation algorithm

To achieve the above max-min fairness criterion, a max-min fair bandwidth allocation scheme (MMFBA) is proposed in [21][9]. For each incoming burst, the arrival rate of the connection which the burst belongs to is estimated using an exponential moving average method. Then LAUC-VF scheduling algorithm [10] is used to

[9] Portions reprinted, with permission, from (Y. Liu, K.C. Chua, and G. Mohan, "Achieving Max-Min Fairness in WDM Optical Burst Switching Networks," in *Proc. IEEE Workshop on High Performance Switching and Routing*, 2005, pp. 187-191). © [2005] IEEE

search the available channels for the incoming burst. If there is an available channel, the burst is scheduled. If not, the algorithm constructs a contention list for the incoming burst. For each channel, only one burst which contends with the incoming burst is put into the contention list. The following steps in the algorithm then determine which burst in the contention list is preempted to achieve max-min fairness. There are four stages in the algorithm.

Stage 1: Determine the max-min fair rate for each connection

Numerous methods can be used to determine such a max-min fair rate given the input rate of each connection. In the proposed scheme, the method in [22] is adopted to calculate the max-min fair rate for each input connection. Since in OBS networks bursts with a larger offset time will have priority over bursts with a smaller offset time, the offset time difference can be converted to a weight ratio when determining the fair rate for each connection. Here, all bursts are assumed to have the same offset time.

Stage 2: Determine the effective load for each connection

In OBS networks, since the link capacity cannot be fully utilized, the max-min fair rate F_i of each connection needs to be converted to a corresponding fair loss probability P_i and max-min fairness provided in this manner. To do this, an effective load E_i is defined for each connection in a reference system with the same link capacity as the actual system. E_i is defined by

$$E_i = \sum_{j=1}^{j=N} \min\{F_i, F_j\}. \tag{5.28}$$

where N is the number of connections sharing a link.

Stage 3: Determine the theoretical loss level for each connection

To determine the loss level for each connection i, the effective load of connection i is fed to the reference system with the same link capacity. Assuming that all traffic arrival patterns are Poisson, the corresponding loss level P_i is then determined by the Erlang formula:

$$P_i = Er(E_i) = \frac{(E_i)^k / k!}{\sum_{n=0}^{n=k} (E_i)^n / n!} \tag{5.29}$$

For those connections with smaller sending rates than their fair rates, the corresponding loss level will be obtained directly by the Erlang formula above. For those connections with larger sending rates than their fair rates, only the loss level for $\frac{F_i}{A_i}$ of the traffic needs to be guaranteed to achieve max-min fairness, where A_i is the measured sending rate of connection i. The loss rate for connection i is thus determined by

$$P_i = \frac{A_i - F_i * (1 - Er(E_i))}{A_i} \tag{5.30}$$

In addition, it is important to ensure that all the loss levels determined in the manner above in the reference system can be guaranteed in the actual system. For connection i, $F_i = A_i$, $A_1 \leq A_2 \leq \cdots \leq A_N$ and $F_1 \leq F_2 \leq \cdots \leq F_N$:

$$\begin{aligned} P_i &= Er(E_i) \\ &= Er\left(\sum_{j=1}^{j=N} \min\{F_i, F_j\}\right) \\ &= Er\left(\sum_{j=1}^{j=i} F_j + (N-i) F_i\right) \end{aligned}$$

The overall loss rate needed to be guaranteed in the reference system is

$$\hat{L} = \sum_{i=1}^{i=N} A_i P_i = \sum_{i=1}^{i=N} A_i Er\left(\sum_{j=1}^{j=i} F_j + (N-i) F_i\right)$$

On the other hand, the overall loss rate in the actual system is

$$L = \sum_{i=1}^{i=N} A_i Er\left(\sum_{j=1}^{j=N} A_j\right)$$

Since for all i, $F_i = A_i$ and $A_1 \leq A_2 \leq \cdots \leq A_N$,

$$\sum_{j=1}^{j=i} F_j + (N-i)\,F_i \leq \sum_{j=1}^{j=N} A_j$$

Since $Er(\cdot)$ is an increasing function,

$$\hat{L} \leq L$$

The above means that not all loss levels in the reference system can be guaranteed in the actual system using the loss level allocation scheme proposed above. To solve this problem, the loss level for the connection with the largest sending rate is increased. The actual loss rate L_i for each connection i is then measured using the moving average method, and the loss level for the connection l with the largest sending rate is determined as follows

$$P_l = \frac{\sum_{i=1}^{i=N} A_i L_i - \sum_{i=1, i\neq l}^{i=N} A_i P_i}{A_l} \tag{5.31}$$

Stage 4. Maintain the loss level for each connection

In this stage, the method in [17] is adopted to maintain the actual loss probability for each connection around its fair loss level determined above. The burst which belongs to the connection whose actual measured loss rate has the largest difference with its loss level determined above will be preempted. A NOTIFY packet will be sent to downstream nodes to remove the reservation of the preempted burst.

Simulation results

The performance of MMFBA is investigated via a simulation study. There are three connections sharing one link with normalized capacity 1. All bursts have the same constant burst length and offset time, and the burst arrival process is Poisson. In the first experiment when the simulation time is less than 0.5s, connections 1, 2 and 3 have normalized sending rates of 0.2, 0.3 and 0.4, respectively. When the simulation time is greater than 0.5s, connection 3 increases its normalized sending rate to 0.6, while the sending rates of connections 1 and 2 remain unchanged. Figure 5.27 shows the observed loss probabilities for the three connections when MMFBA is not employed. In this case, the burst which cannot find an available channel is dropped. After connection 3 increases its sending rate from 0.4 to 0.6, the loss probabilities for connections 1 and 2 also increase to the same level as connection 3 even though connections 1 and 2 do not increase their sending rates. Thus connections 1 and 2 are unfairly affected by connection 3 increasing its sending rate. Figure 5.28 shows the observed loss probabilities for the three connections when MMFBA is used to control the bandwidth allocation. It can be seen from this figure that after connection 3 increases its sending rate from 0.4 to 0.6, the loss probabilities for connections 1 and 2 remain around their theoretical loss levels, thus verifying that MMFBA can provide max-min fairness with good isolation/protection properties.

The second experiment tests the ability of MMFBA in fairly allocating surplus bandwidth. When the simulation time is less than 0.5s, connections 1, 2 and 3 have the same normalized sending rates as those in the first experiment. When the simulation time is greater than 0.5s, connection 1 decreases its normalized sending rate to 0.15, while the sending rates of connections 2 and 3 remain unchanged. Figure 5.29 shows the observed loss probabilities for the three connections when MMFBA is not employed. From this figure, after connection 1 decreases its sending rate to 0.15, the loss probabilities for connections 2 and 3 also decrease to the same level as connection 1. There is not much difference among the loss probabilities experienced by the three connections. Figure 5.30 shows the observed loss probabilities for the three connections when MMFBA is used to control the bandwidth allocation. It can be seen from this figure that after connection 1 decreases its sending rate to 0.15, the loss probabilities for connections 2 and 3 decrease to different levels as determined by their corresponding theoretical loss levels, which verifies that MMFBA can allocate surplus bandwidth in a max-min fair manner in OBS networks.

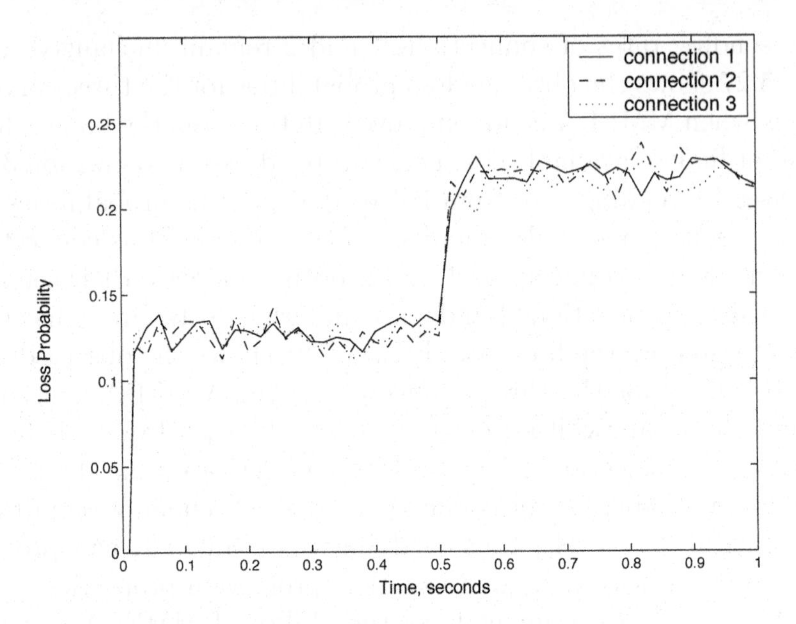

Fig. 5.27. Loss probabilities without max-min bandwidth allocation when connection 3 increases its sending rate at $t = 0.5s$

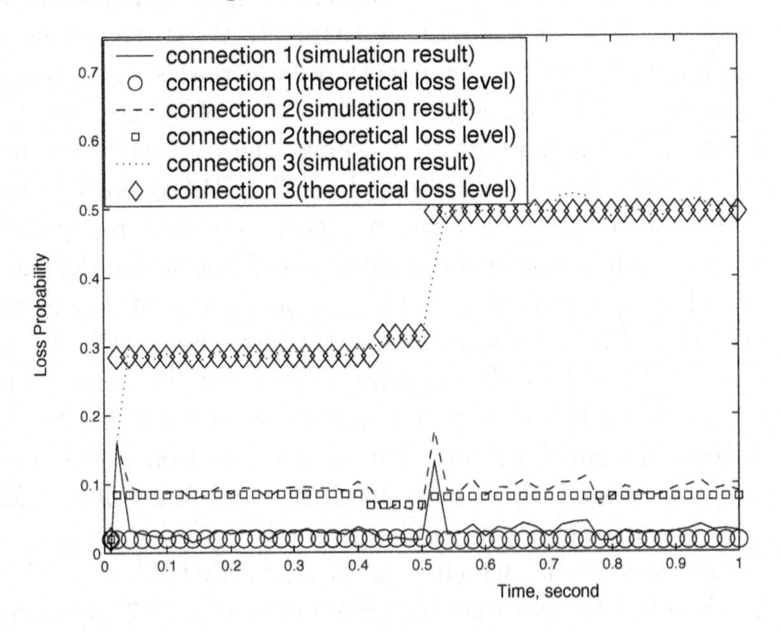

Fig. 5.28. Loss probabilities with max-min bandwidth allocation when connection 3 increases its sending rate at $t = 0.5s$

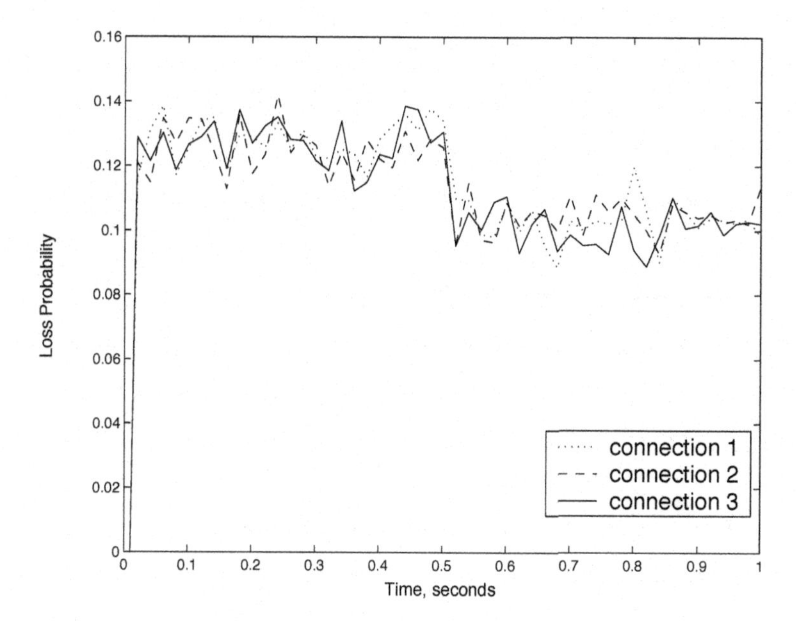

Fig. 5.29. Loss probabilities without max-min bandwidth allocation when connection 1 decreases its sending rate at $t = 0.5s$

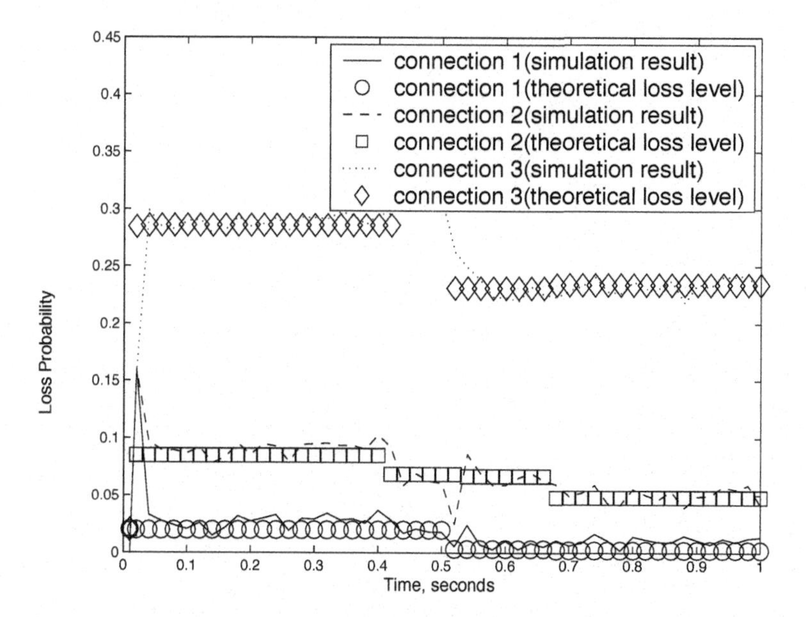

Fig. 5.30. Loss probabilities with max-min bandwidth allocation when connection 1 decreases its sending rate at $t = 0.5s$

References

1. H. Øverby and N. Stol, "Quality of Service in Asynchronous Bufferless Optical Packet Switched Networks," *Telecommunication Systems*, vol. 27, no. 2–4, pp. 151–179, 2004.
2. Q. Zhang, V. M. Vokkarane, J. Jue, and B. Chen, "Absolute QoS Differentiation in Optical Burst-Switched Networks," *IEEE Journal on Selected Areas in Communications*, vol. 22, no. 9, pp. 1781–1795, 2004.
3. M. H. Phùng, K. C. Chua, G. Mohan, M. Motani, and T. C. Wong, "Absolute QoS Signalling and Reservation in Optical Burst-Switched Networks," in *Proc. IEEE Globecom*, 2004, pp. 2009-2013.
4. ——, "An Absolute QoS Framework for Loss Guarantees in OBS Networks," IEEE Transactions on Communications, to appear.
5. E. Rosen, A. Viswanathan, and R. Callon, "Multiprotocol Label Switching Architecture," RFC 3031, 2001.
6. H. Wen, H. Song, L. Li, and S. Wang, "Load-Balancing Contention Resolution in LOBS Based on GMPLS," in *Proc. 4th International Conference on Parallel and Distributed Computing, Applications and Technologies*, 2003, pp. 590–594.
7. J. Zhang, S. Wang, K. Zhu, D. Datta, Y. C. Kim, and B. Mukherjee, "Preplanned Global Rerouting for Fault Management in Labeled Optical Burst-Switched WDM Networks," in *Proc. IEEE Globecom*, 2004. pp. 2004-2008.
8. G. P. Thodime, V. M. Vokkarane, and J. P. Jue, "Dynamic Congestion-based Load Balanced Routing in Optical Burst-Switched Networks," in *Proc. IEEE Globecom*, 2003, pp. 2628-2632.
9. J. Li, G. Mohan, and K. C. Chua, "Dynamic Load Balancing in IP-over-WDM Optical Burst Switching Networks," *Computer Networks*, vol. 47, no. 3, pp. 393–408, 2005.
10. Y. Xiong, M. Vandenhoute, and H. C. Cankaya, "Control Architecture in Optical Burst-Switched WDM Networks," *IEEE Journal on Selected Areas in Communications*, vol. 18, no. 10, pp. 1838–1851, 2000.
11. S. Jamin, S. Shenker, and P. Danzig, "Comparison of Measurement-Based Admission Control Algorithms for Controlled-Load Service," in *Proc. IEEE Infocom*, 1997.
12. M. H. Phùng, K. C. Chua, G. Mohan, M. Motani, and T. C. Wong, "The Streamline Effect in OBS Networks and Its Application in Load Balancing," in *Proc. 2nd International Conference on Broadband Networks*, 2005, pp. 283–290.

13. M. H. Phùng, D. Shan, K. C. Chua, and G. Mohan, "Performance Analysis of a Bufferless OBS Node Considering the Streamline Effect," *IEEE Communications Letters*, vol. 10, no. 4, pp. 293-295, 2006.
14. Q. Chen, G. Mohan, and K. C. Chua, "Offline Route Optimization Considering Streamline Effect in Optical Burst Switching Networks," in *Proc. IEEE International Conference on Communications*, 2006.
15. J. Teng, and G. N. Rouskas, "Routing Path Optimization in Optical Burst Switched Networks," in *Proc. 9th Conference on Optical Network Design and Modeling*, 2005, pp. 1–10.
16. J. Cao, W. S. Cleveland, D. Lin, and D. X. Sun, "The Effect of Statistical Multiplexing on the Long-Range Dependence of Internet Packet Traffic," Bell Labs, Tech. Rep., 2002. [Online]. Available: http://cm.bell-labs.com/cm/ms/departments/sia/doc/multiplex.pdf
17. M. H. Phùng, K. C. Chua, G. Mohan, M. Motani, and T. C. Wong, "A Preemptive Differentiation Scheme for Absolute Loss Guarantees in OBS Networks," in *Proc. IASTED International Conference on Optical Communication Systems and Networks*, 2004, pp. 876–881.
18. G. Mohan, K. Akash, and M. Ashish, "Efficient Techniques for Improved QoS Performance in WDM Optical Burst Switched Networks," *Computer Communications Journal*, vol. 28, no. 7, pp 754-764, 2005.
19. S. K. Tan, G. Mohan and K. C. Chua, "Link Scheduling State Information Based Offset Management in WDM Optical Burst Switching Networks," *Computer Networks Journal*,vol. 45, no. 6, pp 819-834, 2004.
20. B. Zhou, M. Bassiouni and G.Li, "Improving Fairness in optical burst switching networks," Journal of Optical Networking," vol. 3, no.4, pp.214-228, 2004.
21. Y. Liu, K.C. Chua and G. Mohan, "Achieving Max-Min Fairness in WDM Optical Burst Switching Networks", in *Proc. IEEE Workshop on High Performance Switching and Routing*, 2005, pp. 187-191.
22. K. K. Ramakrishnan, D.M. Chiu and R. Jain, "Congestion Avoidance in Computer Networks with a Connectionless Network Layer, Part IV-A Selective Binary Feedback Scheme for General Topologies", DEC Technical Report TR-510, Digital Equipment Corporation, 1987.

6

VARIANTS OF OBS AND RESEARCH DIRECTIONS

This chapter discusses some variants of OBS, including time-slotted OBS, wavelength routed OBS and OBS ring networks. It concludes with a brief description of some possible research issues in OBS networks.

6.1 Time-Slotted OBS

6.1.1 Time-sliced OBS

A time-slotted OBS architecture called Time-Sliced OBS is proposed in [1] to replace switching in the wavelength domain with switching in the time domain. The objective is to eliminate the use of expensive wavelength converters in OBS networks. Figure 6.1 shows the proposed node architecture. There is a *synchronizer* at each incoming link to align the slot boundaries on the link to those of the switch fabric. The synchronizers are followed by FDLs to delay incoming data bursts to provide time-domain switching within a certain range.

In a Time-Sliced OBS network, each wavelength on each link is divided into periodic frames, each of which is further sub-divided into a number of time slots as shown in Figure 6.2. There is a guard time between every time slot. At an ingress node, a data burst is divided into several segments, each of one slot duration. Thus, the burst length is measured in terms of the number of time slots occupied. To simplify the burst switching operation, data

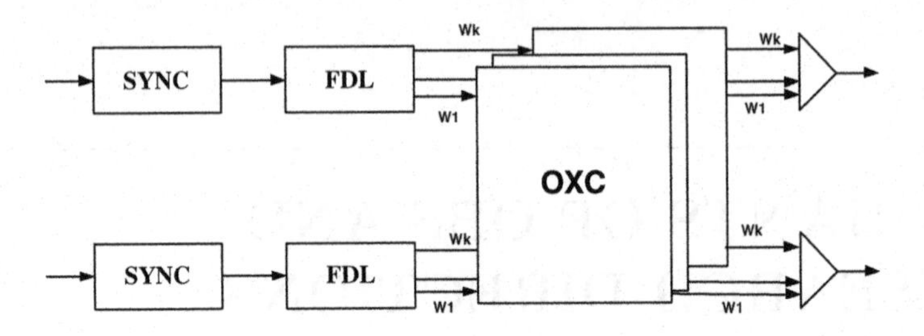

Fig. 6.1. Node architecture for Time-Sliced OBS

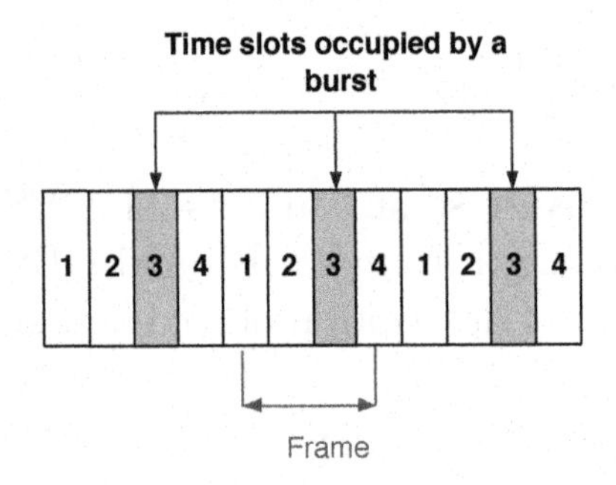

Fig. 6.2. Frame structure for time-sliced OBS

segments from a single burst use the same time slot in a frame for a number of consecutive frames as shown in Figure 6.2. Each burst uses only one time slot within one frame. For example, if the burst length is L time slots, this burst will use the same time slot in each of the L consecutive frames.

The header packet of a burst indicates the arrival time of the first data segment of the burst, the time slot position within a frame and the number of time slots the burst uses. If all the required time slots are available, the burst is scheduled and the corresponding time slots are reserved. Otherwise, the scheduler delays the burst using FDLs to await available time slots. To limit the size of the FDLs, the maximum delay is chosen to be the number of time slots in one frame. It is noted that the amount of delay for each data segment of the same burst is the same in order to keep all the segments at the same time slot within a frame. If the available time slots cannot be found within the maximum delay, the burst is dropped.

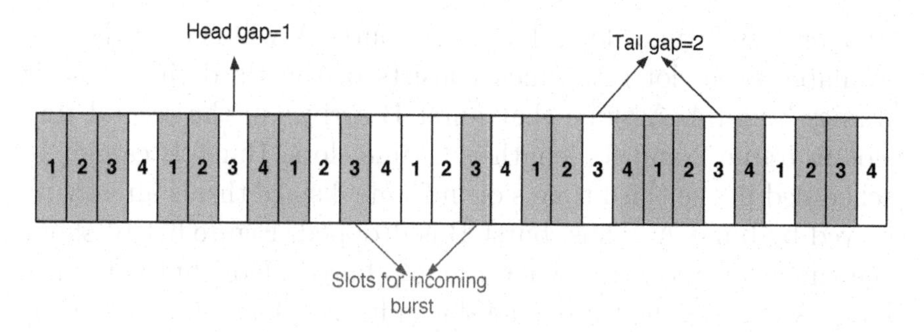

Fig. 6.3. Gap definition of Time-sliced OBS

In [2], another burst scheduling algorithm is proposed to improve the burst loss probability in Time-Sliced OBS. The burst scheduler at each node scans all available time slot sets for each incoming burst and chooses the time slot set that can minimize the gaps between the newly reserved burst and existing bursts. Thus, the probability that future bursts can find available time slot sets can be improved. If each time slot position within a frame is considered a separate channel, the gap-based scheduling algorithm for Time-Sliced OBS is similar to LAUC-VF, which has been described in Chapter 2 for traditional OBS.

As shown in Figure 6.3, for a given time slot position, a gap is defined as the number of frames that the time slot position is unoccupied between two different consecutive bursts that are assigned to the time slot. If an incoming burst is shorter than the gap, it can be successfully scheduled. The gap before the first time slot of a burst is defined as the head gap, while the gap after the last time slot of the same burst is defined as the tail gap. Each burst is associated with one head gap and one tail gap for each available time slot set. For example, Figure 6.3 shows an incoming burst of length 2 time slots. The head gap for this incoming burst is 1 while the tail gap is 2. The burst scheduler selects the available time slot set that minimizes either the head gap or the tail gap.

Figure 6.4 illustrates the advantage of the gap-based algorithm [2]. In Figure 6.4(a), the header packet for burst A arrives earlier than that for burst B. Therefore, burst A is scheduled first. Assume that burst A arrives at the second time slot in frame 3 and its length is 2 time slots. Therefore, burst A will choose the first available time slot set, which consists of the third time slots in frames 3 and 4. Assume that burst B arrives in the second time slot in frame 2 and its length is 2 time slots. Burst B cannot be scheduled in the third time slot in frame 3 since this is already reserved by burst A. Thus, burst B is dropped. Figure 6.4(b) shows the time slot allocation when the gap-based algorithm is used. If burst A chooses the third time slot in frames 3 and 4, its head gap is 1 and its tail gap is 2. However, if it chooses the fourth time slot in frames 3 and 4, its head and tail gaps will be 0. Since the burst scheduler selects the available time slot set for which either the head gap or the tail gap is the minimum, the fourth time slot in frames 3 and 4 will be allocated to burst A. In this case, burst B is successfully scheduled in the third time slot in frames 2 and 3.

6.1.2 Optical burst chain switching

One-way reservation is used in [1, 2]. As a result, burst collisions happen within core nodes resulting in burst loss. As in WDM circuit switching networks, burst collisions can be avoided by having explicit wavelength reservation before data is transmitted. However, having explicit reservation for each burst will result in excessive signaling overheads. In [3], an *Optical Burst Chain Switching* (OBCS) mechanism, which combines the merits of optical circuit switching and optical burst switching, is proposed. In OBCS, the switching entity is a *burst chain,* which consists of multiple non-contiguous and non-periodic bursts on a wavelength. Collision-free burst transmission is achieved by reserving bandwidth on an e2e basis for each burst chain in a time-slotted optical network.

Figure 6.5 shows the burst chain building process of OBCS. A source node continuously measures the burst arrival rate of each connection between itself and every destination node. At the start of a frame, the source node uses the average burst arrival rate measured in the last frame as the predicted bandwidth demand. The value is attached to a PROBE packet, which is sent towards the destination node to collect the time slot availability information at

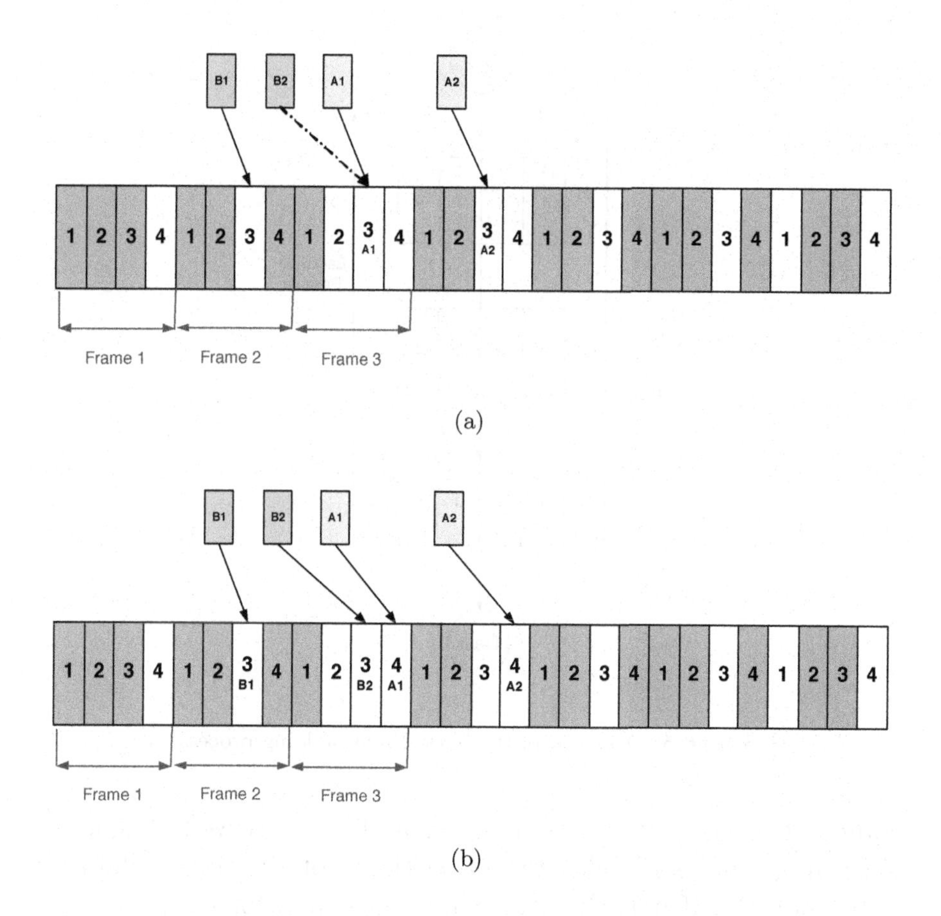

Fig. 6.4. Burst scheduling in Time-Sliced OBS

each core node along the path. When the PROBE packet reaches the destination node, the node starts to search for the required number of time slots according to the information collected by the PROBE packet. At this time, the PROBE packet has an array of the time slot availability information for each core node along the path as shown in Figure 6.6. Once a slot is found to be available, the algorithm will check whether the reserved bandwidth exceeds the predicted bandwidth requirement.

This searching algorithm is illustrated using an example as shown in Figure 6.6. Here, if time slot 1 at core node R1 is available and the reserved bandwidth does not exceed the predicted

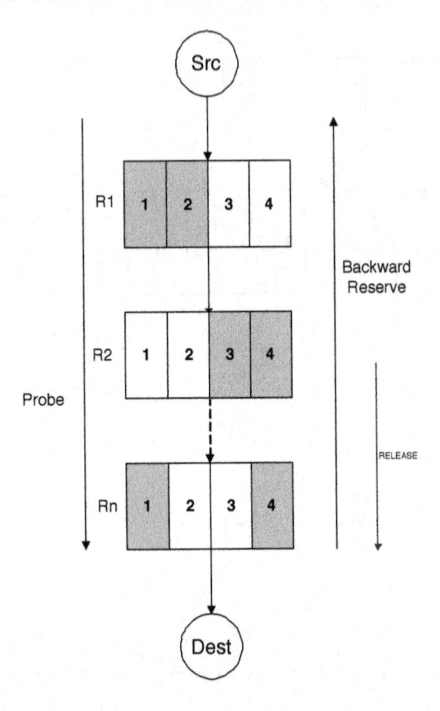

Fig. 6.5. Overview of the burst chain building process

bandwidth requirement, time slot 1 at R1 is reserved. Then, the searching process will check whether time slot 1 is also available at core node R2. If not, the algorithm will search for the first available time slot within a feasible range (determined by the maximum number of FDLs provided at each wavelength) at this node. If the algorithm can find a free time slot (e.g. time slot 3) at R2 then this time slot is reserved and the same process is repeated for the next core node (R3), etc. In this way, a series of time slots (1→3→4) are set up for the connection. This series of time slots is referred to as a time slot light path vector.

Because of FDL limitations, some time slots that are available may not be usable. For example, time slot 4 in R1 is available. However, if the maximum number of FDLs available, which determines the feasible delay range for a node at a core node, supports a maximum delay of 4 time slots, no feasible slot in R2 can be found for time slot 4 in R1. In this case, the unfinished light path vector originating from time slot 4 in R1 is released and the next

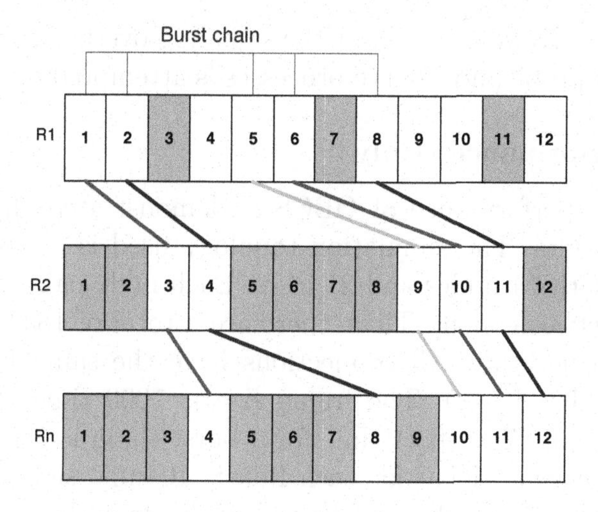

Fig. 6.6. Time slot searching process

available time slot in R1 is selected to begin another time slot searching process for feasible time slots in R2 and R3 to form the light path vector. Thus, in the figure, several light path vectors are set up by the PROBE packet. These can be used to build a *burst chain* at R1 (time slots 1, 2, 5, 6, 8) that will be switched as a whole entity at R2 (time slots 3, 4, 9, 10, 11) and at R3 (time slots 4, 8, 10, 11, 12). Note, however, that these time slot light path vectors still need to be reserved before they can be used to switch a burst chain.

To reserve the time slots for a burst chain, a RESERVE packet is sent back by the egress node towards the ingress node. At each intermediate node, the chosen time slots are reserved if they remain available[1]. Otherwise, the reservation process fails and all the time slots reserved at core nodes up until the current node need to be released. A RELEASE packet is sent from the current node towards the destination node for this purpose.

If enough time slots needed to meet the requirement of a burst chain cannot be reserved, another probe and reserve process can be initiated if there is sufficient time before the departure of the

[1] Due to cross traffic, a time slot may be chosen by other burst chains

burst chain. However, to limit the signaling overhead, a maximum number of probe and reserve processes is attempted in each frame.

6.1.3 Performance study

In [3], the performance of OBCS is demonstrated through simulation results. The simulation topology used is a NSFNET-like network with 14 nodes and 21 links. Each node pair has one connection between them. Thus, there are a total of 183 connections inside the network. All connections have the same Poisson distributed offered load. The frame size is 1000 time slots of $1us$ duration each. The offset time for a burst chain is 500 time slots. The propagation delay for each link is 10 time slots. The maximum number for probe and reserve cycles in each frame is set to be 5, and the buffer size is set to be 100 time slots. The number of FDLs provided for each channel is 3.

OBCS is compared against OBS with wavelength converters (OBS-WC) and OBS without wavelength converters (OBS-NWC). The burst size for OBS is fixed at 1 time slot. The number of wavelengths per link is 16.

Figure 6.7 shows the throughput comparison between the three architectures. OBCS is seen to have much better performance than OBS with/without wavelength converters. This is because the time slotted nature of OBCS forces bursts to be aligned and thereby increases channel utilization. On the other hand, the fact that the two OBS architectures are asynchronous increases the chance of burst contention and reduces network throughput. Between the two OBS architectures, OBS with wavelength converters performs better since the presence of wavelength converters allows a better level of statistical multiplexing.

Figure 6.8 shows the effect of the maximum number of probe and reserve cycles in each frame on the performance of OBCS. The offered load is 0.07 and the buffer size is set to be 100 time slots. As expected, the throughput of OBCS increases with the number of probes because more time slots can be potentially reserved with more probes. However, beyond a certain maximum number of probes, the additional improvement is marginal because there is not enough time to probe for more available time slots.

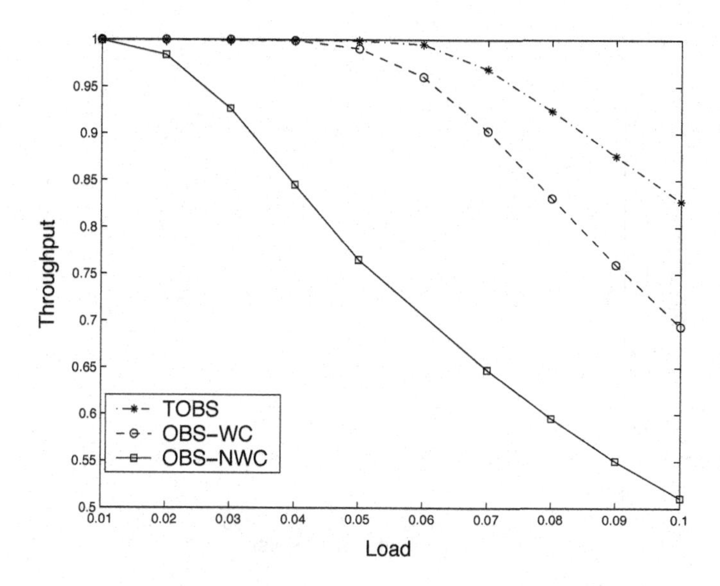

Fig. 6.7. Throughput comparison with varying offered load © [2005] IEEE

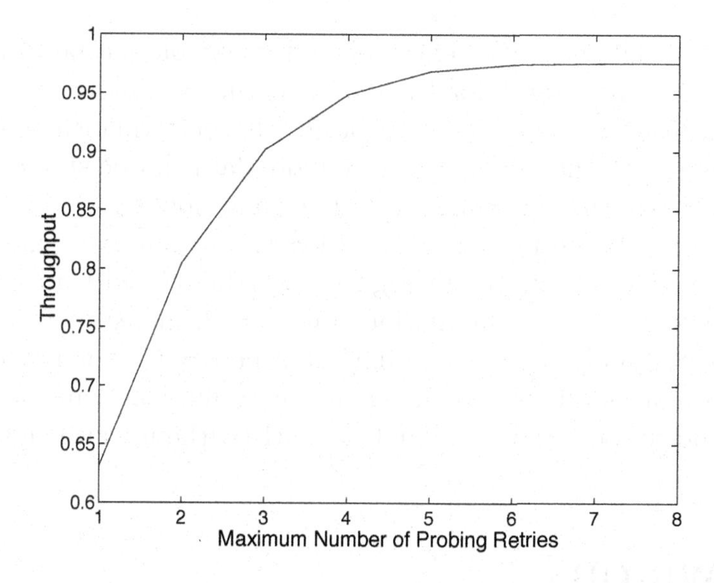

Fig. 6.8. Throughput under different maximum number of probes in each frame © [2005] IEEE

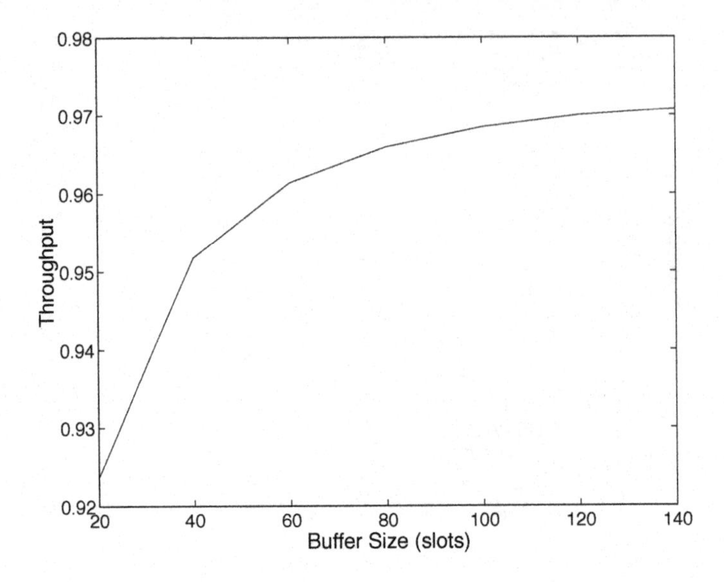

Fig. 6.9. Throughput under different buffer size © [2005] IEEE

Figure 6.9 shows the effect of buffer size on the performance of OBCS. The offered load is 0.07 and the maximum number of probes in each frame is set to be 5. The performance of OBCS increases with the buffer size since more data can be stored in the buffer to wait for transmission in free time slots.

Finally, the performance of OBCS is also observed under dynamic traffic conditions. In Figure 6.10, the offered load for each source is 0.07 before simulation time $t = 2$, increases to 0.09 at time $t = 2$ and goes back to 0.07 at time $t = 6$. It is noted that OBCS adapts quickly to the changing traffic loads and achieves much better performance than OBS with/without wavelength converters.

6.2 WR-OBS

In [4], a two-way resource reservation mechanism called wavelength-routed OBS is proposed to solve the burst collision problem observed in one-way reservation schemes. When a burst is ready at an ingress node, a bandwidth request is sent to a centralized resource allocation controller. The controller searches for any free

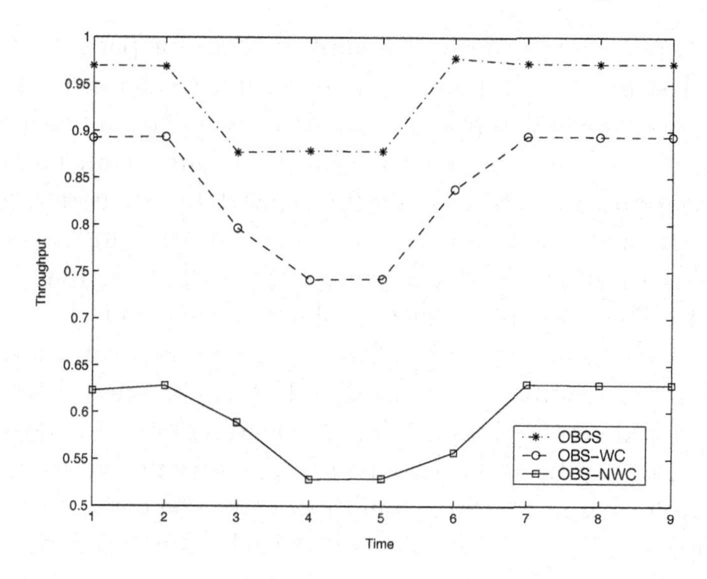

Fig. 6.10. Throughput under dynamic traffic for Time-Slotted OBS networks © [2005] IEEE

wavelength on all paths from the ingress node to the egress node. If a free wavelength is found, an ACK message is sent back to the ingress node allowing the burst to be sent over the selected wavelength. Since the wavelength connection is established before the transmission of each burst, the e2e delay can be guaranteed. However, the scheme suffers from scalability and signaling overhead problems since the centralized controller needs to maintain global wavelength utilization and to process all burst requests generated by all edge nodes in the entire OBS network. In [5], WR-OBS is extended to support differentiated QoS through using differentiated treatment for requests from different traffic classes at the centralized controller.

6.3 OBS in Ring Networks

In [6], several ring access protocols are proposed for a WDM metro ring network with optical burst switching. The unidirectional metro ring network consists of OBS nodes, which are responsible for burst assembly and transmission. In addition, each

OBS node maintains several separate queues for packets with different destinations and some fair queueing mechanisms are used to support QoS differentiation. To reduce burst contention, each node has a dedicated home wavelength for burst transmission.

A common control wavelength is used by all nodes to send their control signaling information. The control information from all nodes are grouped into a control frame that circulates in the network. There is one dedicated slot for each node in a control frame to insert its control information. The control information specifies the destination address, offset time, burst length and other related control information. Depending on the length of the ring, there may be several control frames circulating on the control wavelength in the ring simultaneously. The control frame will be dropped by each node and added back after inserting its own control information if it needs to send bursts. Each node tunes its receiver to the specified wavelength at the specified time according to the information obtained from the control frame. There are only two pairs of transceivers at each node. One pair is fixed, which is used by the control wavelength to transmit and receive the control frames. The other pair is used by the home wavelength to transmit data bursts. The receiver for this pair is not fixed and can be tuned to different wavelengths dynamically to receive data bursts. Although there is no burst contention on the transmission wavelengths, there could be contention at the receiver side due to receiver conflict since it may so happen that multiple bursts arrive at a destination node simultaneously. To reduce such burst contentions, the following access protocols have been proposed [6].

6.3.1 Round-robin with random selection (RR/R)

In this protocol, each OBS node serves the waiting queues in a round robin manner. The source node assembles bursts using packets waiting in the selected queue. The source node monitors the control frames. When there is a burst ready for transmission at time t, the source node inserts the control information on its own slot in the first control frame arriving after time t. Then, the burst is transmitted on its home wavelength after the specified offset time. If the destination node finds that the arrival time of all incoming bursts will overlap, it will randomly choose one burst and the optical receiver will tune to the selected wavelength at the given time. This scheme is fair for the waiting queues in each node. However, some bursts will be lost due to the receiver conflict.

6.3.2 Round-robin with persistent service (RR/P)

This scheme reduces receiver conflicts by monitoring the control information carried by the control frame. Particularly, each node keeps a record of the earliest free time of all possible destination nodes. This record will be updated by each control frame passing through the node. When a source node wants to send a burst, which is assembled by packets from one selected queue to a particular destination node, the source node calculates the arrival time of the burst at that destination node. If the arrival time is found to be later than the earliest free time of that node, the burst will be sent. If the arrival time is found to be earlier than the earliest free time of that node, the burst will not be sent since it would otherwise cause a receiver conflict. The burst will wait for the next control frame. The same process will be repeated until the burst is sent. The scheme is said to use persistent service since the source node will not proceed to serve other queues until the burst from the selected queue is sent. Since this scheme is a distributed scheme, it may happen that two nodes will send bursts to one destination node simultaneously based on the control information collected from different control frames. Therefore, this scheme cannot completely remove burst contentions. The following scheme tries to reduce the end-to-end burst transmission delay by using non-persistent service.

6.3.3 Round-robin with non-persistent service (RR/NP)

This scheme is similar to the round-robin scheme except that it is based on non-persistent service. The difference is that once a source node finds that the outgoing burst will cause receiver conflicts, it proceeds to serve the next waiting queue. Thus, end-to-end burst transmission delay is reduced. However, the scheme could lead to starvation since the source node will always give up serving a burst if it will lead to a contention.

6.3.4 Round-robin with tokens (RR/Token)

This scheme uses multiple tokens to avoid burst contention. The basic idea is to allocate a token for each node and at any time only one token is used by a source node for one destination such that simultaneous transmissions can be avoided. A token has two states: namely free and busy. If node j is free, token j for this node will be set as free in the control frame. In all subsequent control frames, token j will be set as busy. The first node that captures the control frame will remove free tokens from the control frame and put these onto a FIFO queue in the token arrival order. These free tokens will be marked as busy in the current control frame and will be transmitted to all the downstream nodes. The first node will check the queue lengths for the free tokens inside the token queue. If the queue for destination j is not ready (not enough packets to form a burst), token j will be released to the next control frame and token j will be marked as free in the next control frame. If the queue for destination j is ready, the source node will assemble a burst from the packets in the selected queue and the burst will be sent after the offset time. Token j will also be marked as busy in the control frame. Once the burst transmission is finished, the destination node will mark token j as free for the next control frame and the same process will be repeated. Since at any instant of time, only one source holds the free token for any destination node, there will be no burst contention.

6.3.5 Round-robin with acknowledgement (RR/Ack)

This scheme achieves contention-free burst transmission using an acknowledgement mechanism. A source node serves a waiting queue in a round-robin manner. The source node puts the request onto the next control frame. Unlike the other schemes, the request does not specify the offset time. Upon receiving the request, the destination node determines the offset time and sends the acknowledgement back to the source node. The source node is not allowed to send another request before the acknowledgement is received to avoid transmission conflicts.

Based on the simulation results presented in [6], in terms of mean node throughput, RR/Token has the best performance, followed by RR/P, RR/NP, and RR/R. In terms of mean packet delay, RR/R has the smallest mean packet delay, followed by RR/NP, RR/P and RR/Token. RR/ACK has the same performance as RR/Token in terms of mean node delay, but has a longer mean packet delay than RR/Token.

6.4 Optical Burst Transport Networks

Although OBS can provide fine bandwidth granularity and achieve high statistical multiplexing gain, a lot of transmission ports are required for each switch in OBS networks. By considering this problem, a new architecture is proposed in [7] to transport bursts over a virtual topology and to reduce the number of ports required. The proposed scheme is called Optical Burst Transport Network (OBTN). In OBTN, a burst can be transmitted in two modes. The first mode is to transmit bursts on direct end-to-end lightpaths between any two OBTN nodes. The second mode is to transmit burst on hop-by-hop links as individual bursts using multiplexing. The first mode is used with high priority since it can help to reduce operational complexity and number of ports required for each switch. The second mode is only used when necessary as an alternative way to support bursty traffic. Corresponding to the burst transport modes, there are two types of links in OBTN networks. The first one is called *circuit link*, which is the direct lightpath connecting any two OBTN nodes. The second one is called *burst link* where bursts are multiplexed.

6.4.1 OBTN node architecture

The architecture of an OBTN node is shown in Figure 6.11. There are three kinds of links for an OBTN node: links to metro networks, circuit links, hop-by-hop burst links. The transport mode for each burst can be determined end-to-end or locally at each

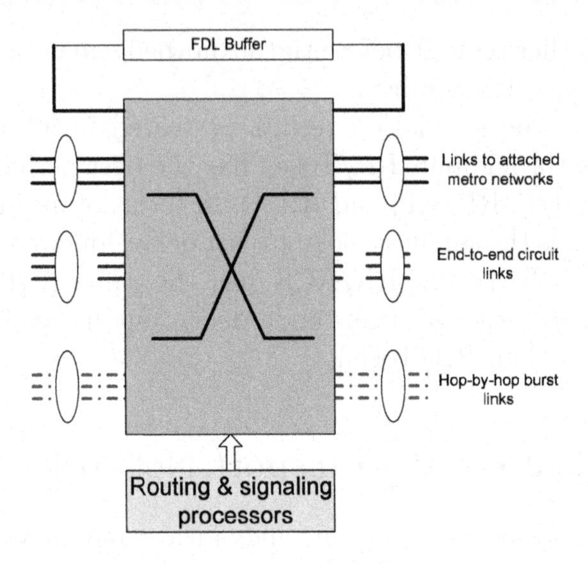

Fig. 6.11. OBTN node architecture

OBTN node. In addition, FDL buffer and wavelength convertors could be employed to achieve high network resource utilization.

6.4.2 OBTN architecture

An OBTN is shown in Figure 6.12. Circuit links are denoted by solid lines and burst links are denoted by dashed lines. As shown, the number of burst links are relatively small compared to the number of circuit links.

The combination of circuit switching and burst switching poses some new challenges for efficient burst transmission over OBTN. These challenges include transport mode selection, routing selection, QoS and survivability issues. Research on OBTN is still in progress.

6.5 Optical Testbed

6.5.1 Optical burst transport ring

A 2.5Gbps optical burst transport ring testbed is described in [8]. The prototype is implemented using Field Programmable Gate

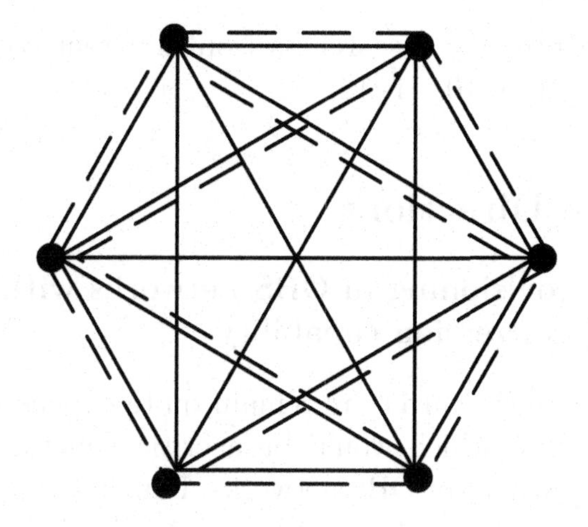

Fig. 6.12. OBTN architecture

Arrays (FPGAs). In this prototype testbed, separate wavelengths are used for control panel and data panel. A token is used for medium access control within the ring. Each data channel has its own controlling token. The token is transmitted and processed in the control plane at every node. A node can start transmission on a data channel only when it grabs a token. The optical switch in the ring has two states: bar state and cross state. In the bar state, all cross traffic will transparently bypass the intermediate node. When a switch is transmitting or receiving data, the switch is in the cross state for a certain amount of time. For data transmission process, the time the switch is in cross state is determined by the duration of the outgoing burst. For data reception, the time the switch is in the cross state is the duration of the incoming burst which is notified in advance by the corresponding header packet for the incoming burst. The header packet has 40 bits for the control message. The transmission speed of the control channel is 1.25 Gbps. The transmission speed of the data channel is 2.5 Gbps. The guard time between consecutive data bursts is limited by the switching configuration time of switching devices. In the prototype, the guard time is set at 256 ns. The switching configuration time is 80 ns. There are also some other OBS testbeds that are developed by researchers from different countries. They are described in [9, 10, 11, 12].

6.6 Future Directions

6.6.1 QoS provisioning in OBS networks with partial wavelength conversion capability

Optical wavelength converters remain quite expensive under current technologies, which would be a critical factor affecting the practical deployment of OBS networks. Partial (or no) wavelength conversion capability poses challenges for QoS provisioning. In view of this, it is an important issue to consider QoS provisioning for OBS networks with partial wavelength conversion capability.

6.6.2 QoS provisioning in time-slotted OBS networks

As has been discussed earlier in this chapter, time-slotted OBS has been proposed as a promising switching architecture for OBS networks since it not only can eliminate the use of expensive wavelength converters, but also can help to reduce burst contention owing to the fact that bursts are transmitted in a synchronous manner. Some preliminary works have been carried out mainly on the basic architecture of time-slotted OBS networks. Due to its special switching architectures, an important direction is to investigate QoS provisioning mechanisms in time-slotted OBS networks.

References

1. J. Ramamirtham and J. Turner, "Time Sliced Optical Burst Switching," in *Proc. Infocom*, 2003, pp. 2030–2038.
2. T. Ito, Daisuke Ishii, K. Okazaki, and I. Sasase, "A Scheduling Algorithm for Reducing Unused Timeslots by Considering Head Gap and Tail Gap in Time Sliced Optical Burst Switched Networks," *IEICE Transactions on Communications*, Vol.J88-B, No.2, pp. 242–252, 2006.
3. Y. Liu, G. Mohan, and K. C. Chua, "A Dyamic Bandwidth Reservation Scheme for a Collision-Free Time-Slotted OBS Network," in *Proc. IEEE/CreateNet Workshop on OBS*, pp. 192–194, 2005.
4. M. Duser and P. Bayvel, "Analysis of a Dynamically Wavelength-Routed Optical Burst Switched Network Architecture," *IEEE/OSA Journal of Lightwave Technology*, vol. 20, no. 4, pp. 574–585, 2002.
5. E. Kozlovski, M. Dueser, A. Zapata, and P. Bayvel, "Service Differentiation in Wavelength-Routed Optical Burst-Switched Networks," in *Proc. IEEE/OSA Conference on Optical Fibre Communications*, pp. 774–775, 2002.
6. L. Xu, H. G. Perros, and G.N. Rouskas, "A Simulation Study of Access Protocols for Optical Burst-Switched Ring Networks," in *Proc. International IFIP-TC6 Networking Conference on Networking Technologies, Services, and Protocols*, 2002, pp. 863–874.
7. C.M. Gauger and B. Mukherjee, "Optical burst transport network (OBTN) - a novel architecture for efficient transport of optical burst data over lambda grids," in *Proc. High Performance Switching and Routing*, pp. 58–62, 2005.
8. J. Kim, et al., "Demonstration of 2.5 Gbps Optical Burst Switched WDM Rings Network," in *Proc. Optical Fiber Communication Conference*, 2006.
9. S. Junghans, "A Testbed for Control Systems of Optical Burst Switching Core Nodes," in *Proc. the Third International Workshop on Optical Burst Switching (WOBS)*, 2004.
10. K. Kitayama, et al., "Optical burst switching network testbed in Japan," in *Proc. Optical Fiber Communication Conference*, 2005.
11. Y. Sun, T. Hashiguchi, V. Minh, X. Wang, H. Morikawa, and T. Aoyama, "A Burst Switched Photonic Network Testbed: Its Architecture, Protocols and Experiments," *IEICE Transactions on Communications*, vol. E88-B, no. 10, pp. 3864-3873, 2005.

12. H. Guo, et al., "Design and Implementation of Edge Node in TBOBS," In *Proc. SPIE Network Architectures, Management, and Applications II*, vol. 5626, 2005.
13. D. Griffith and S. K. Lee, "A 1+1 Protection Architecture for Optical Burst Switched Networks," *IEEE Journal on Selected Areas in Communications*, vol. 21, no. 9, pp. 1384–1398, 2003.

Index